R. J. GRAY
80.04.09

Fibre Cements
and
Fibre Concretes

Fibre Cements and Fibre Concretes

D. J. Hannant

Department of Civil Engineering
University of Surrey

A Wiley—Interscience Publication

JOHN WILEY & SONS
Chichester · New York · Brisbane · Toronto

Copyright © 1978, by John Wiley & Sons, Ltd.

All rights reserved.

No part of this book may be reproduced by any means, nor transmitted, nor translated into a machine language without the written permission of the publisher.

Library of Congress Cataloging in Publication Data:
Hannant, D. J.
    Fibre cements and fibre concretes.
    'A Wiley–Interscience publication.'
    Includes index.
    1. Reinforced concrete, Fiber. 2. Fibers.
3. Cement – Additives. I. Title.
TA444.H33   666'.8   77-28323
ISBN 0 471 99620 3

Typeset by Preface Ltd, Salisbury, Wilts
and printed by Unwin Brothers Ltd.,
The Gresham Press, Old Woking.

Acknowledgements

In a work such as this on a developing subject, one is inevitably dependent on the advice of experts in specific topics. In this respect I am greatly indebted to Mr. R. I. T. Williams of the University of Surrey for contributing Chapter 12 on road and airfield pavements, to Mr. J. J. Zonsveld of the University of Surrey for his assistance and advice regarding Chapter 7 on polypropylene fibres, to Dr. A. J. Majumdar of the Building Research Establishment for reading and commenting on Chapter 8 on glass fibres and to Mr. J. Aveston of the National Physical Laboratory for reading Chapter 3 on theory.

In addition, many others have assisted with the work, either by contributing material or reading parts of the manuscript.

The following Crown copyright figures and tables appear by courtesy of the Director, Building Research Establishment, and are published by permission of the Controller of Her Majesty's Stationery Office:

Figures 3.11, 5.1, 5.3, 6.4, 8.3, 8.7, 8.8, 8.9, 8.10, 8.11, 8.12, 8.13, 8.15, 8.16, 8.17, 8.18, 8.19, 8.21, 8.23, 8.24, 8.25, 8.26, 8.29, 8.30, 8.31, 9.4; and Tables 6.1, 8.2, 8.3, 8.4, 8.5.

Contents

List of Symbols ix

Foreword xi

Preface xiii

1. Introduction 1

2. Properties of materials 3

3. Theoretical principles of fibre reinforcement in uniaxial tension 8

4. Theoretical principles of fibre reinforcement in flexure 35

5. Steel-fibre concrete: Properties in the fresh state and mix design for workability 52

6. Steel-fibre concrete: Properties in the hardened state 61

7. Polypropylene fibres in concrete, mortar, and cement 81

8. Glass fibres in cement and in concrete 99

9. Asbestos cement 135

10. Fibres other than asbestos, glass, polypropylene, and steel 146

11. Applications of steel, polypropylene, glass, and asbestos fibres 157

12. Steel-fibre concrete in road and airfield pavements 182

Appendix 1. Theory for minimum crack spacing (x') of long, non-circular fibres with frictional bond 198

Appendix 2. Practical examples using the theoretical treatment for typical real composites 200

Solutions to practical examples 203

Author Index 211

Subject Index 215

List of Symbols

A	cross-sectional area
A_c, A_f	cross-sectional area of composite, fibre
A_g	$\dfrac{\text{weight of aggregate greater than 5 mm}}{\text{total weight of concrete}}$
D	depth of beam section
D_f, D_m	density of fibre, matrix
E	modulus of elasticity
E_c, E_f, E_m	modulus of elasticity of composite, fibre, matrix
F	force in tension
M	modular ratio
N	number of fibres crossing unit area of composite
P	load in bending
P_f	perimeter of fibre
SFS	specific fibre surface, i.e. total surface area of all fibres within a unit volume of composite
V	volume
V_c, V_f, V_m	volume of composite, fibre, matrix
$V_{f\,(\text{crit})}$	volume of fibres which, after matrix cracking, will carry the load which the composite sustained before cracking
V_f'	effective volume of fibre in direction of stress calculated from V_f using efficiency factors
W_f	weight of fibres as a percentage of total composite weight
W_f'	weight of fibres as a percentage of weight of matrix
c	composite
d	fibre diameter
d_n	distance of neutral axis from tensile face of beam
f	fibre
l	fibre length
l_c	critical fibre-length i.e. twice the length of fibre embedment which would cause fibre failure in a pull-out test
l_a	lever arm of beam in bending
m	matrix

r	fibre radius
t	tensile
u	ultimate in tension
w	crack width
x'	transfer length for stress for long fibres (Also minimum crack spacing)
x''	final average crack spacing
x_d	transfer length for stress for short fibres
α	Scaling factor $= \dfrac{E_m V_m}{E_f V_f}$
ϵ	strain
$\epsilon_c, \epsilon_f, \epsilon_m$	strain in composite, fibre, matrix
ϵ_{fu}	ultimate strain in fibre
$\epsilon_{comp}, \epsilon_t$	compressive strain, tensile strain
ϵ_x	strain at start of major multiple cracking
η_1	efficiency factor for fibre orientation
η_2	efficiency factor for fibre length
σ	stress
$\sigma_c, \sigma_f, \sigma_m$	stress in composite, fibre, matrix
σ_{comp}	comprehensive stress in composite
σ_{cu}	ultimate, post cracking tensile strength of composite
σ_{fu}	failure stress of fully bonded fibres or pull-out stress of debonded fibres
σ_{mu}	ultimate tensile strength of matrix, (i.e. cracking stress)
σ_{MR}	modulus of rupture of composite
σ_t	cracking stress of composite
τ	average sliding friction bond strength
τ_d	bond strength after fibre slip
τ_s	bond strength before fibre slip

Foreword

In this book Dr. Hannant puts together a description of the science and of the technology of fibre cements and fibre concretes. The incorporation of fibres into brittle things has been an elementary technology for millenia. The writer of the Book of Exodus refers to the impossibility of making bricks without straw (Exodus, Chapter 5, verse 6, *et seq.*) and the technology of asbestos reinforced cement has been known for about seventy years. However, the scientific principles behind the understanding of how fibres incorporated into brittle things prevent these breaking, has only recently been explored, understood, and rationally applied. The spur to do this came from the advances in the glass and textile industries and from the search for light and stiff materials for use in aerospace. Because of these very diverse sources, a book is needed which brings together results from many parts of the scientific and technical literature. Dr Hannant has contributed to the scientific understanding himself and in this work gives a comprehensive and fully up-to-date treatment designed for use by the engineer and by the scientist.

By using the effects described in this book, cement can be fashioned into shapes and forms hitherto reserved for much more expensive and less common materials. Cement, an artificial stone, first produced, we believe, by the Romans and rediscovered in the late eighteenth century, can now, by incorporation of fibres, be made in certain circumstances almost unbreakable.

As the demands on our energy resources grow, the use of materials such as cement, which cost little in energy to produce compared with steel or the plastics, enable our resources to be stretched still further. The energy requirement to produce a cubic metre of steel is 22 times that necessary to produce a corresponding volume of cement and the energy cost of cement per unit volume is less than one-fifth that of the common plastics. Further the minerals required to make cement are present on our earth in almost limitless quantities and geographically widely distributed. These two features lend impetus to the incorporation of fibres into cement as an advantage for civilization generally.

A. Kelly, Sc.D., F.R.S.,
Vice-Chancellor University of Surrey

Preface

There is an increasing tendency towards specialization in many branches of science and engineering but in the field of fibre cement and fibre concrete a variety of disciplines have combined in an attempt to produce pseudo-ductile materials from fibres embedded in relatively brittle cement based matrices.

Physicists, mathematicians, materials scientists, chemists, and engineers have all contributed to the development of the subject and the resulting research publications are widely scattered in journals and conference proceedings and are not necessarily in a form which is readily usable by the practising engineer or by the materials scientist.

In writing this book, therefore, I have attempted to bridge the gap between the research worker and those who wish to put the new materials to practical use so that the latter may have a comprehensive text covering properties, production, and applications of a wide range of fibre cements and fibre concretes. The basic theoretical principles are outlined and simplified theoretical treatments are given which are nevertheless of sufficient accuracy to be used for many practical situations.

The subject is a developing one and not all the materials described herein will ultimately prove successful but I hope that a sound knowledge of the basic principles governing the physical behaviour of the composites will encourage their use in appropriate situations.

D.J.H.

Chapter 1
Introduction

The properties of unreinforced cement and concrete have been well documented[1-4] and an understanding of the basic principles of concrete technology has been assumed in writing the text.

One of the problems of a cement-based matrix is the inherently brittle type of failure which occurs under tensile stress systems or impact loading and in the construction industry, a major reason for the growing interest in the performance of fibres in cement based materials is the desire to increase the toughness or tensile properties of the basic matrix.

Emphasis on energy conservation has also stimulated interest in methods of replacing materials such as cast iron, glass-reinforced plastics, and bituminous materials by the use of fibre cements and fibre concretes. This idea of replacing existing materials also extends into the asbestos-cement industry, where, due to problems relating to the supply of asbestos in the long-term and also because of possible health hazards, attention is being given to the replacement of asbestos with man-made fibres.

In order to be able to satisfy the performance requirements of these various interests, adequate material properties must be achieved in the fibre composite and the main objectives of the materials engineer in attempting to modify the properties of cement or concrete are as follows:

(a) To improve the tensile or flexural strength;
(b) To improve the impact strength;
(c) To control cracking and the mode of failure by means of post-cracking ductility;
(d) To change the rheology or flow characteristics of the material in the fresh state.

There is no doubt that these objectives can be achieved in the short term but a degree of caution is necessary for most fibre concretes and cements regarding the long term performance, which, in the Civil Engineering sense may well be in excess of 50 years. This is likely to result in the bulk of applications being 'non-structural' in that no great danger to life should result from component collapse in the event of unforseen deterioration with time.

REFERENCES

1. Neville, A. M., *Properties of Concrete*, Pitman, 1977.
2. Orchard, D. F., *Concrete Technology*, Volumes 1 and 2, John Wiley and Sons Inc., 1968.
3. Troxell, G. E., Davis, H. E., and Kelly, J. W. *Composition and Properties of Concrete*, McGraw-Hill Book Co., 1968.
4. Czernin, W., *Cement Chemistry and Physics for Civil Engineers*, Crosby Lockwood and Son Ltd., 1970.

Chapter 2
Properties of Materials

2.1 PROPERTIES OF FIBRES AND OF MATRICES

The main factors controlling the theoretical performance of the composite material are the physical properties of the fibres and the matrix and the strength of the bond between the two. Values for bond are given in the appropriate sections with qualifying remarks because bond strengths vary with a wide variety of ill-defined parameters, including time. Typical ranges for the other physical properties of the fibres are shown in Table 2.1 and the properties of some matrices are shown in Table 2.2.

It is apparent from these tables that the elongations at break of all the fibres are two or three orders of magnitude greater than the strain at failure of the matrix and hence the matrix will crack long before the fibre strength is approached. This fact is the reason for the emphasis on post-cracking performance in the theoretical treatment.

On the other hand, the modulus of elasticity of the fibre is generally less than five times that of the matrix and this, combined with the low fibre volume fraction, means that the modulus of the composite is not greatly different from that of the matrix.

The fibre types in Table 2.1 can be divided into two main groups, these with moduli lower than the cement matrix, such as cellulose, nylon and polypropylene and those with higher moduli such as asbestos, glass, steel, carbon, and Kevlar, which is a form of aromatic polyamide introduced by DuPont. The last two are included for the sake of completeness but high cost would seem to rule them out for major engineering applications.

The low modulus organic fibres are generally subject to relatively high creep which means that if they are used to support permanent high stresses in a cracked composite, considerable elongations or deflections may occur over a period of time. They are therefore more likely to be used in situations where the matrix is expected to be uncracked, but where transitory overloads such as handling stresses, impacts or wind loads are significant.

Another problem with the low modulus fibres is that they generally have large values of Poisson's ratio and this, combined with their low moduli, means that if

Table 2.1. Typical fibre properties

Fibre	Diameter μm	Length mm	Density Kg/m³ 10³	Young's Modulus GN/m²	Poisson's Ratio	Tensile Strength MN/m²	[a]Elongation at break %	Typical Volume in composite %
Asbestos								
Chrysotile (white)	0.02–30	≤40	2.55	164	0.3	200–1800 (fibre bundles)	2–3	10
Crocidolite (blue)	0.1–20	—	3.37	196	—	3,500	2–3	—
Carbon								
Type 1 (High modulus)	8	10-continuous	1.90	380	0.35	1,800	~0.5	2–12
Type 2 (High strength)	9		1.90	230		2,600	~1.0	
Cellulose			1.2	10		300–500		10–20
Glass Cem-Fil filament								
E	8–10	10–50	2.54	72	0.25	3,500	4.8	2–8
	12.5			80	0.22	2,500	3.6	
204 filament strand	110 × 650		2.7	70	—	1,250	—	
Kevlar								
PRD 49	10	6–65	1.45	133	0.32	2,900	2.1	<2
PRD 29	12		1.44	69	—	2,900	4.0	
Nylon (Type 242)	>4	5–50	1.14	Rate dependent up to 4	0.40	750–900	13.5	0.1–6
Monofilament	100–200	5–50	0.9	Rate dependent up to 5	—	400	18	0.1–6
Polypropylene					0.29—			
Fibrillated	500–4000	20–75	0.9	up to 8	0.46	400	8	0.2–1.2
Steel								
High tensile	100–600	10–60	7.86	200		700–2000	3.5	0.5–2
Stainless	10–330			160	0.28	2,100	3	

[a]Note: one per cent elongation = 10,000 × 10⁻⁶ strain

Table 2.2. Typical properties of matrix

Matrix	Density Kg/m^3	Youngs Modulus GN/m^2	Tensile Strength MN/m^2	Strain at failure x 10^{-6}
Ordinary Portland Cement Paste	2,000–2,200	10–25	3–6	100–500
High Alumina Cement Paste	2,100–2,300	10–25	3–7	100–500
O.P.C. Mortar	2,200–2,300	25–35	2–4	50–150
O.P.C. Concrete	2,300–2,450	30–40	1–4	50–150

stretched along their axis, they contract sideways much more than the other fibres. This leads to a high lateral tensile stress at the fibre-matrix interface which is likely to cause a short aligned fibre to debond and pull out. Devices such as woven meshes or networks of fibrillated fibres may therefore be necessary to give efficient composites.

Even the high modulus short fibres may require mechanical bonding to avoid pull out unless the specific surface area is very large. Thus steel fibres are commonly produced with varying cross-sections or bent ends to provide anchorage and glass fibre bundles may be penetrated with cement hydration products to give an effective mechanical bond after a period of time.

Four types of matrix are shown in Table 2.2 and the two cement types shown are important mainly because they have different degrees of alkalinity which affect the durability of glass and steel fibres.

The maximum particle size of the matrix is also important because it affects the fibre distribution and the quantity of fibres which can be included in the composite. The average particle size of cement paste before hydration is between 10 and 30 microns whereas mortar is considered to contain aggregate particles up to 5 mm maximum size. Concrete which is intended to be used in conjuction with fibres should not have particles greater than 20 mm and preferably not greater than 10 mm otherwise uniform fibre distribution becomes difficult to achieve.

In order to avoid shrinkage and surface crazing problems in finished products it is advisable to use at least 50 per cent by volume of inert mineral filler, which may be aggregate or could include pulverized fuel ash, or limestone dust. However, if the inert filler consists of a large volume of coarse aggregate the volume of fibres which can be included will be limited which will in turn limit the tensile strength and ductility of the composite.

Strength of the matrix is mainly affected by the free water/cement ratio and this parameter also has a lesser effect on the modulus so that the properties of the matrices shown in Table 2.2 can vary widely.

The values for fibre density in Table 2.1 are important in that they enable the relationship between fibre volume, which is required for the theoretical treatment, and fibre weight, which is required for batching, to be calculated. This calculation is complicated by the problem that in practice, fibre weight (W_f') is often expressed as a proportion of the weight of the cement or concrete matrix whereas in theoretical treatments, fibre weight (W_f) and fibre volume (V_f) are generally expressed as proportions of the whole composite weight or volume.

For example:

(a) In practice
$$W_f' = \frac{\text{Weight of fibre}}{\text{Weight of matrix}} \times 100 \text{ per cent}$$

$$= \frac{V_f D_f}{V_m D_m} \times 100 \text{ per cent} \tag{2.1}$$

(b) In theoretical work
$$W_f = \frac{\text{Weight of fibre}}{\text{Weight of matrix} + \text{Weight of fibre}} \times 100 \text{ per cent}$$

$$= \frac{V_f D_f}{V_m D_m + V_f D_f} \times 100 \text{ per cent} \tag{2.2}$$

Using equations (2.1) and (2.2) for steel fibres at 2 per cent by volume ($V_f = 0.02$) in concrete together with the material properties shown in Tables 2.1 and 2.2 it follows that:

In equation (2.1)
$$W_f' = \frac{0.02 \times 7860}{0.98 \times 2350} \times 100 = 6.84 \text{ per cent}$$

In equation (2.2)
$$W_f = \frac{0.02 \times 7860}{0.98 \times 2350 + 0.02 \times 7860} \times 100 = 6.42 \text{ per cent}$$

Thus, the same volume of steel fibres (2 per cent) in the same matrix may be quoted either as 6.84 per cent or 6.42 per cent by weight depending on the requirements for batching or theoretical analysis, respectively. Similar considerations apply to other fibre types.

The commercial viability of fibre composites is critically dependent on the costs of the fibres which exercise a controlling influence on the cost of the product because the matrix is so cheap.

Thus, a commercial decision on whether an existing product can be profitably replaced with one made from fibre cement or fibre concrete may depend on using the cheapest suitable fibre in the lowest volume required to fulfil the strength and durability requirement.

Unfortunately it is not possible to give specific prices because costs of fibres are subject to relatively rapid changes depending on demand and energy costs in their production. However, the cost of fibre per cubic metre of composite can be calculated from equation (2.3) provided that the fibre cost in pence/kg is known.

Cost of fibre, in £ per m^3
of composite = V_f × fibre cost in pence/kg × 10 × fibre specific gravity   (2.3)

For example a polypropylene fibre composite containing 6 per cent by volume of fibres costing 100 p/kg would result in a cost of fibres in the composite of 0.06 × 150 × 10 × 0.9 £/m^3 i.e. £54/m^3.

Similarly, 6 per cent by volume of glass fibres costing 150p/kg would cost 0.06 × 150 × 10 × 2.7 £/m^3, i.e. £243/m^3 would be the cost of fibres alone.

It is apparent from these crude calculations that some careful commercial decisions may have to be taken regarding savings in labour or transport costs compared with the costs of equivalent products in timber, steel, aluminium, plastic, or reinforced concrete before these new materials are used in large quantities.

2.2 PROPERTIES OF THE COMPOSITE MATERIALS

The properties of the composite materials are detailed in the appropriate chapters but the reader should be aware that the quoted properties are often highly dependent on the test techniques used to measure them.

For instance, in impact testing, it is well known that quoted values for 'impact strength' or 'toughness' are often meaningless because the recorded value is so dependent on factors such as the energy and velocity of the impacting mass, the size of specimen and rigidity of supports, the type of test and even the definition of failure.

It is less well known that the slow or 'static' techniques developed for testing unreinforced concrete can give similarly misleading information if they are used to obtain material descriptors for quasi-ductile fibre reinforced composites. This is because unreinforced concrete is relatively brittle and is often regarded in International Standard Methods of Test as linearly elastic up to failure. Although this may be a relatively acceptable approximation for plain concrete, it is often inappropriate for fibre concretes where the post-cracking ductility may form a major part of the load carrying system. Thus, indirect methods of test for tensile strength such as split cylinder or modulus of rupture in bending can lead to overestimates of the tensile strength as measured by direct tests by more than 100 per cent.

Appropriate test techniques for fibre cements and fibre concretes have not yet been agreed by the major International Standards Institutes and therefore, in the following text, emphasis is often placed on the limitations of the test techniques which have been used to determine the material properties.

Chapter 3
Theoretical Principles of Fibre Reinforcement in Uniaxial Tension

3.1 GENERAL BACKGROUND

Much of the original theoretical work on the properties of fibre composites was carried out for fibre-resin systems in which the fibre volume often exceeded 50 per cent and the fibre properties were used to increase both the stiffness and the strength of the resin in order that stiff, lightweight, high tensile strength components could be produced for the aero-space industry. However, the low volume of fibres and the relatively high stiffness of the cement or concrete matrix have lead to a change of emphasis in the theoretical treatment of fibre cements and concretes because the increase in ductility after cracking is a major factor to be considered in the utilization of these materials.

Cement paste and concrete have certain limitations with respect to fibre reinforcement, and the main features must be appreciated before the mechanics of fibre reinforcement can be placed on a sound theoretical basis.

The major limitations are as follows:

(a) Low tensile failure strain ($< 500 \times 10^{-6}$).
(b) Relatively high modulus of elasticity ($7-40$ GN/m^2) which, although useful in limiting deflections in structural situations, results in relatively little use being made of the load carrying capacity of the fibres until after cracking has occurred.
(c) Limited capacity for the incorporation of fibres. (Generally less than 10 per cent by volume in cement paste and less than 2 per cent by volume in concrete which may already contain 70 per cent of its volume as aggregate particles.)
(d) Relatively poor bonding with many fibre types.
(e) High alkalinity of the paste (pH 12–13). This protects steel fibres but can cause glass fibres to deteriorate with time.

It should be noted that (a), (b), and (c) above result in less efficient fibre incorporation as we move from cement paste to concrete.

Rigorous theoretical treatments of fibre strengthening mechanisms are available[1,2,3] but for cement based composites, where the matrix failure strain is much less than that of the fibres, the fibre volume rarely exceeds 8 per cent, the distribution of fibres may be between two-dimensional or random three-dimensional and the fibres short with relatively poor bond, many of the refinements of these theoretical treatments are inappropriate for the prediction of material properties. Other factors which tend to prevent accurate prediction of composite properties are the time dependent variables such as the static and sliding friction bond strengths between the fibres and the matrix. These critical parameters are known to vary as the matrix continues to hydrate but precise strengths at a given time are generally a matter for conjecture.

Nevertheless, it is worth examining simplified theories in order to be able to predict the 'order' of possible improvements to the matrix.

A major difficulty has to be resolved before a theoretical treatment can be related to measured material properties and this is the meaning of the commonly used term 'first crack strength'. Cracks can be detected by eye or by sophisticated electronic techniques and the loads at which cracking is said to occur can differ by more than 100 per cent depending on the technique used. Strictly, 'first crack' should apply to the stage at which microscopic parts of the paste or paste and aggregate separate but these micro-cracks can be stable even in a direct tensile stress system without fibre reinforcement.

For the purposes of the following theoretical analysis, therefore, cracking is considered to be at the rather indeterminate stage where the cracks normally start to propagate just before visible cracking occurs. This stage is referred to in the analysis as the ultimate strength of the matrix.

3.2 SIMPLIFIED THEORY FOR UNIAXIAL TENSION BEFORE CRACKING

The following theory has been given in most text books on composites[1,2,3] and is commonly referred to as the 'laws of mixtures' by materials scientists although the principles are the same as those used in reinforced concrete theory. None of the assumptions is likely to be true in practice.

Assumptions:

(i) The fibres are aligned in the direction of the stress
(ii) Before cracking the fibres are fully bonded to the matrix, i.e. equal strains in fibre and matrix
(iii) The Poisson's ratio in fibre and matrix = 0

Definitions

A – area; V – volume; E – modulus of elasticity; σ – stress; ϵ – strain; F – force.

Suffixes

f – fibre; m – matrix; c – composite; u – ultimate.

Let cross-sectional area of composite $A_c = 1$

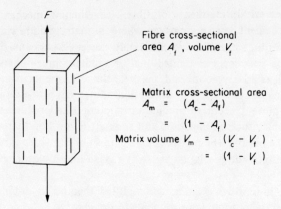

Figure 3.1. Aligned fibre composite tested in uniaxial tension

Let volume of composite $\quad V_c = 1$

Fibre volume, V_f, is expressed as a fraction of the volume of the composite, i.e.

$$\frac{V_f}{V_c} = \frac{V_f}{1}$$

$$\epsilon_c = \epsilon_f = \epsilon_m = \frac{\sigma_c}{E_c} = \frac{\sigma_f}{E_f} = \frac{\sigma_m}{E_m} \tag{3.1}$$

$$F = \sigma_c A_c = \sigma_f A_f + \sigma_m A_m$$

∴ The average stress (σ_c) carried by a unit area of composite ($A_c = 1$) at a given strain (ϵ_c) before cracking is given by

$$\sigma_c = \sigma_f A_f + \sigma_m (1 - A_f)$$

In fibre technology we usually use volume fractions, and for unit length, $V_c = A_c = 1$, and $V_f = A_f$

i.e. $\sigma_c = \sigma_f V_f + \sigma_m (1 - V_f)$ \hfill (3.2)

$$\therefore \quad \frac{\sigma_c}{\sigma_m} = \frac{\sigma_f}{\sigma_m} V_f + 1 - V_f$$

From (3.1)

$$\frac{\sigma_f}{\sigma_m} = \frac{E_f}{E_m} = M \quad \text{the modular ratio}$$

$$\therefore \quad \frac{\sigma_c}{\sigma_m} = 1 + V_f(M - 1) \tag{3.3}$$

$$\therefore \quad \sigma_m = \frac{\sigma_c}{1 + V_f(M-1)} \tag{3.4}$$

$$\frac{\sigma_c}{\sigma_m} = \frac{E_c}{E_m} = 1 + V_f\left(\frac{E_f}{E_m} - 1\right)$$

$$\therefore \quad E_c = E_f V_f + E_m(1 - V_f) \tag{3.5}$$

From equation (3.4) the effect of the inclusion of fibres on the stress in the composite at which the matrix cracks can be determined as follows:

(a) Steel-fibre concrete

$$V_f = 0.02 \quad E_f = 200 \text{ GN/m}^2 \quad E_m = 30 \text{ GN/m}^2$$

from equation (3.4)

$$\sigma_m = \frac{\sigma_c}{1 + 0.02(6.7 - 1)} = \frac{\sigma_c}{1.11}$$

(b) Glass-reinforced cement

$$V_f = 0.05 \quad E_f = 70 \text{ GN/m}^2 \quad E_m = 17 \text{ GN/m}^2$$

$$\sigma_m = \frac{\sigma_c}{1 + 0.05(4.12 - 1)} = \frac{\sigma_c}{1.16}$$

Assuming that the matrix cracking stress is not altered by the presence of fibres then it is apparent that even with a relatively large volume of high-modulus fibres, fully bonded and aligned in the most favourable direction, the cracking stress in the composite is only increased by about 11 per cent and 16 per cent compared with the unreinforced matrix for (a) and (b) respectively. This increase could be achieved more cheaply in other ways, such as by a reduction in the w/c ratio.

Also, because the modular ratio (M) is less than one for low modulus fibres, equation (3.4) implies that the composite will crack at an increasingly lower stress than the matrix alone as the volume of low modulus fibres increases.

The change in modulus of the composite due to fibre addition can be determined for the same two materials from equation (3.5), i.e.

(a) Steel-fibre concrete

$$E_c = 200 \times 10^9 \times 0.02 + 30(0.98)10^9$$
$$= 33.4 \text{ GN/m}^2$$

(b) Glass-reinforced cement

$$E_c = 70 \times 10^9 \times 0.05 + 17(0.95)$$
$$= 19.6 \text{ GN/m}^2$$

It can be shown from these calculations that with the most favourable assumptions, the *pre-cracking* performance, either for increased stiffness or cracking stress, is not likely to be greatly improved by the addition of fibres. In fact, Allen[4] has shown that even with glass fibres the initial composite modulus can be less than that of the matrix if allowance is made for the various efficiency factors described in Sections 3.2.1 and 3.2.2.

3.2.1 Fibre Orientation

The orientation of the fibres relative to the direction of stress can be allowed for by the substitution of $\eta_1 V_f$ instead of V_f in the first term of equations (3.2) and (3.5), where η_1 is an efficiency factor.

The value of η_1 for a given fibre orientation depends on the method of analysis used but some typical values are given in Table 3.1.

Table 3.1. Efficiency factors, η_1, for a given fibre orientation relative to the direction of stress

Fibre orientation		η_1 according to Cox[5]	Krenchel[6]
1-D	aligned	1	1
2-D	random in plane	$\frac{1}{3}$	$\frac{3}{8}$
3-D	random	$\frac{1}{6}$	$\frac{1}{5}$

It is apparent from Table 3.1 that non-aligned fibres are likely to have much less effect than aligned fibres on the material properties under direct stress in the uncracked state.

3.2.2 Fibre length

In the region before the matrix failure strain is reached Laws[7] considers that the length efficiency factor for an aligned short fibre composite with frictional bond at the fibre interface, is nearly unity (i.e. 0.98) for practical composites.

However Allen[4] considers that for thin composites, a second efficiency factor (η_2) is necessary where (η_2) depends on fibre length. Alternative forms of equation (3.2) and (3.5) are therefore:

$$\sigma_c = \eta_1 \eta_2 \sigma_f V_f + \sigma_m (1 - V_f) \tag{3.6}$$

$$E_c = \eta_1 \eta_2 E_f V_f + E_m (1 - V_f) \tag{3.7}$$

The value of (η_2) is determined by the length of the fibres (l) in relation to the critical fibre length 'l_c' which is defined as twice the length of fibre embedment which would cause fibre failure in a pull out test.

Allen[4] has given the following values for η_2.

$$l \leqslant l_c \quad \eta_2 = \frac{l}{2l_c} \tag{3.8}$$

$$l \geqslant l_c \quad \eta_2 = 1 - \frac{l_c}{2l} \qquad (3.9)$$

In practical terms, however, there is probably little point in worrying about the precise value for 'η_2' in the calculation of the composite cracking stress or pre-cracking modulus because the critical fibre length is rarely known with any accuracy and will probably vary with age at test as the bond strength increases with time. Also, σ_c and E_c are generally so close to the matrix values that second order theoretical effects are likely to be swamped by the variability in properties which is normally experienced with cement paste and concrete.

3.3 SIMPLIFIED THEORY TO CALCULATE CRITICAL FIBRE VOLUME $V_{f(crit)}$

It has already been shown that it is unlikely to be very beneficial to include fibres in cement matrices to increase the cracking stress, and therefore, the merit if any, of fibre inclusion must lie in the load carrying ability of the fibres after matrix cracking has occurred.

The cracked composite may carry a lesser or greater load after cracking than the uncracked material as shown in Figures 3.2(a) and 3.2(b). Both curves may have practical value although multiple cracking is more likely to occur for the material described by Figure 3.2(b) and this endows the composite with greater apparent ductility.

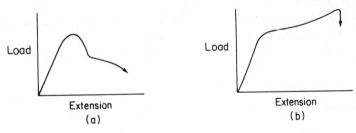

Figure 3.2. Typical load—extension curves in direct tension

In order that the material may follow the curve in Figure 3.2(b), it is necessary for the critical fibre volume $V_{f(crit)}$ to be exceeded.

The critical fibre volume is defined as the volume of fibres which, after matrix cracking, will carry the load which the composite sustained before cracking.

Using the rather unrealistic assumptions given in Section 3.2 $V_{f(crit)}$ can be determined as follows:

Let ϵ_{mu} = matrix cracking strain

σ_{mu} = matrix cracking stress

$V_{f(crit)}$ = critical volume of fibres

σ_{fu} = maximum failure stress of fully bonded fibres, or pull out stress of debonded fibres

At cracking

$$\epsilon_f = \epsilon_{mu}$$

$$\sigma_f = \epsilon_{mu} \cdot E_f$$

$$\sigma_{mu} = \epsilon_{mu} \cdot E_m$$

Substituting in equation (3.2), for the situation just before cracking

$$\sigma_c = \epsilon_{mu} E_f V_{f(crit)} + \sigma_{mu}(1 - V_{f(crit)}) \tag{3.10}$$

After cracking $\sigma_{mu} = 0$, and for the same σ_c to be carried by the fibres alone:

$$V_{f(crit)} \sigma_{fu} = \epsilon_{mu} E_f V_{f(crit)} + \sigma_{mu}(1 - V_{f(crit)}) \tag{3.11}$$

$$\therefore \quad V_{f(crit)} = \frac{\sigma_{mu}}{(\sigma_{fu} - \epsilon_{mu} E_f + \sigma_{mu})} \tag{3.12}$$

Equation (3.12) can also be expressed as

$$V_{f(crit)} = \frac{E_c \epsilon_{mu}}{\sigma_{fu}} \tag{3.13}$$

where E_c is defined by equation (3.5)

For economic reasons $V_{f(crit)}$ should be as low as possible. Equation (3.12) is mainly controlled by the terms σ_{mu} and σ_{fu} because generally $\sigma_{fu} \gg \sigma_{mu}$ and the implications from the equation are therefore as follows:

Factors most likely to reduce $V_{f(crit)}$:

1. Reduction in σ_{mu} (1st order). This implies that if multiple cracking is to be achieved a low matrix strength might be useful.
2. Increase in σ_{fu} (1st order). This implies an increase in pull out load of de-bonded fibres.
3. Reduction in ϵ_{mu} or E_f (2nd order).

Factors most likely to increase $V_{f(crit)}$.

1. Random fibres (1st order).
2. Poor bond (1st order).

3.3.1 Graphical Representation of Critical Fibre Volume $V_{f(crit)}$

A graphical representation of the position of $V_{f(crit)}$ in relation to equation (3.2) is shown in Figure 3.3.

It can be seen from Figure 3.3 that the slope of the line $\sigma_{fu} V_f$ greatly affects the intersection point at which the critical fibre volume for strengthening in direct tension occurs, and hence for fibres which generally pull out at failure the bond strength is the controlling influence on $V_{f(crit)}$. It is also apparent from the figure that the strength of the composite, where the critical volume fraction is exceeded is given by $\sigma_{fu} V_f$ where σ_{fu} is the stress to either break or pull out the fibres.

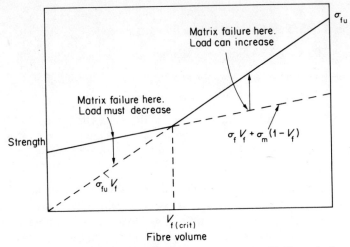

Figure 3.3. Graphical representaion of equation (3.2) and the position of critical fibre volume

3.3.2 Idealized Calculations to find the Critical Fibre Volume ($V_{f(crit)}$) for Strengthening in Direct Tension

Assuming the most favourable factors of fully bonded, continuous aligned fibres the minimum critical fibre volumes can be calculated for typical concretes and pastes as follows:

Example 1. Steel fibres in concrete

$\epsilon_{mu} = 100 \times 10^{-6}$ $\sigma_{mu} = 3 \text{ MN/m}^2$

$E_f = 200 \text{ GN/m}^2$ $\sigma_{fu} = 1000 \text{ MN/m}^2$

From equation (3.12)

$$V_{f(crit)} = \frac{3 \times 10^6}{1000 \times 10^{-6} - 100 \times 10^{-6} \times 200 \times 10^9 + 3 \times 10^6}$$

$$= \frac{3}{983} \simeq 0.31 \text{ per cent}$$

Example 2. Glass fibre bundles in cement paste

$\epsilon_{mu} = 300 \times 10^{-6}$ $\sigma_{mu} = 5 \text{ MN/m}^2$

$E_f = 70 \text{ GN/m}^2$ $\sigma_{fu} = 1250 \text{ MN/m}^2$

The value for σ_{fu} for glass fibre bundles was taken from References 8 and 9

$$V_{f(crit)} = \frac{5 \times 10^6}{1250 \times 10^6 - 300 \times 10^{-6} \times 70 \times 10^9 + 5 \times 10^6}$$

$$= \frac{5}{1234} \simeq 0.4 \text{ per cent}$$

Example 3. Fibrillated polypropylene fibres in concrete

$\epsilon_{mu} = 100 \times 10^{-6}$ $\sigma_{mu} = 3 \text{ MN/m}^2$

$E_f = 8 \text{ GN/m}^2$ $\sigma_{fu} = 400 \text{ MN/m}^2$

$$V_{f(crit)} = \frac{3 \times 10^6}{400 \times 10^6 - 100 \times 10^{-6} \times 8 \times 10^9 + 3 \times 10^6}$$

$$= \frac{3}{402.2} \simeq 0.75 \text{ per cent}$$

Although the fibre volumes calculated in Examples 1, 2, and 3 can be economically included in cement paste or concrete, when these idealised examples are examined in realistic terms, it can be shown that the critical fibre volumes are likely to be increased to values which may be commercially unattractive and from practical manufacturing aspects very difficult to include in the matrix.

3.4 STRESS–STRAIN CURVE, MULTIPLE CRACKING, AND ULTIMATE STRENGTH

If the critical fibre volume for strengthening has been reached then it is possible to achieve multiple cracking of the matrix. This is a desirable situation because it changes a basically brittle material with a single fracture surface and low energy requirement to fracture, into a pseudo-ductile material which can absorb transient minor overloads and shocks with little visible damage. The aim of the materials engineer is often therefore to produce a large number of cracks at as close a spacing as possible so that the crack widths are very small (say < 0.1 mm). These cracks are almost invisible to the naked eye in a rough concrete surface and the small width reduces the rate at which aggressive materials can penetrate the matrix when compared with commonly allowable widths in reinforced concrete of up to 0.3 mm.

High bond strength helps to give a close crack spacing but it is also essential that the fibres de-bond sufficiently local to the crack to give ductility which will absorb impacts.

Aveston *et al.*[10,11,12] have given a clear description of the principles behind the calculation of the complete stress strain curve, the crack spacing and the crack width for long and short aligned fibres for the simplified case where the bond between the fibres and matrix is purely frictional and the matrix has a well defined single valued breaking stress. Sections 3.4.1 and 3.4.3 are based on Reference 12 by permission of the Director of the National Physical Laboratory, Teddington, Middlesex. (Crown copyright reserved.)

The theory in Section 3.4.1 has been developed for circular fibres but the same principles apply to fibres of irregular cross section and comparable equations for non-circular fibres are given in Appendix 1.

3.4.1. Long fibres with frictional bond

The idealized stress–strain curve for a fibre reinforced brittle matrix composite is shown in Figure 3.4.

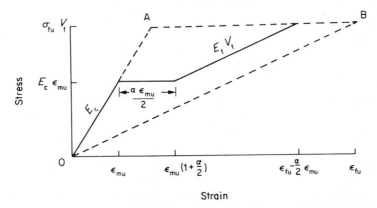

Figure 3.4. Idealized stress–strain diagram for a brittle matrix composite (Reproduced from Aveston, Mercer, and Sillwood,[12] *National Physical Laboratory Report No. SI*, No. 90/11/98 Part I, January 1975, by permission of the National Physical Laboratory. Crown copyright reserved)

The initial Young's modulus (E_c) has already been given by equation (3.5) and the derivation of the rest of the curve on Figure 3.4 is as follows.

If the fibre diameter is not too small, the matrix will fail at its normal failure strain (ϵ_{mu}) and the subsequent behaviour will depend on whether the fibres can withstand the additional load without breaking, i.e. whether

$$\sigma_{fu} V_f > E_c \epsilon_{mu} \tag{3.13}$$

If they can take this additional load it will be transferred back into the matrix over a transfer length (x') (Figure 3.5) and the matrix will eventually be broken down into a series of blocks of length between (x') and ($2x'$).

We can calculate (x') from a simple balance of the load ($\sigma_{mu} V_m$) needed to break unit area of matrix and the load carried by N fibres of radius (r) across the same area after cracking. This load is transferred over a distance (x') by the limiting maximum shear stress (τ).

i.e. $\quad N = V_f / \pi r^2$

$\quad\quad 2\pi r N \tau x' = \sigma_{mu} V_m$

or

$$x' = \frac{V_m}{V_f} \cdot \frac{\sigma_{mu} r}{2\tau} \tag{3.14}$$

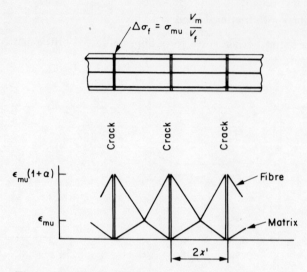

Figure 3.5. Strain distribution after cracking of an aligned brittle matrix composite (Reproduced from Aveston, Mercer, and Sillwood,[12] *National Physical Laboratory Report No. SI,* No. 90/11/98 Part I, January 1975, by permission of the National Physical Laboratory. Crown copyright reserved)

The stress distribution in the fibres and matrix (crack spacing $2x'$) will then be as shown in Figure 3.5.

The additional stress ($\Delta\sigma_f$) on the fibres due to cracking of the matrix varies between $\sigma_{mu}V_m/V_f$ at the crack and zero at distance x' from the crack so that the average additional strain in the fibres, which is equal to the extension per unit length of composite at constant stress $E_c\epsilon_{mu}$ is given by

$$\Delta\epsilon_c = \tfrac{1}{2}\sigma_{mu} \cdot \frac{V_m}{V_f} \cdot \frac{1}{E_f}$$

i.e.
$$\Delta\epsilon_c = \frac{\epsilon_{mu}E_m V_m}{2E_f V_f} = \frac{\alpha\epsilon_{mu}}{2} \tag{3.15}$$

where

$$\alpha = E_m V_m / E_f V_f \tag{3.16}$$

and the crack width, (w) bearing in mind that the matrix strain relaxes from ϵ_{mu} to $\epsilon_{mu}/2$ will be given by

$$w = 2x'\left(\frac{\alpha\epsilon_{mu}}{2} + \frac{\epsilon_{mu}}{2}\right)$$

or

$$w = \epsilon_{mu}(1 + \alpha)x' \qquad (3.17)$$

At the completion of cracking the blocks of matrix will all be less than the length ($2x'$) required to transfer their breaking load ($\sigma_{mu}V_m$) and so further increase in load on the composite results in the fibres sliding relative to the matrix, and the tangent modulus becomes ($E_f V_f$).

In this condition, the load is supported entirely by the fibres and the ultimate strength (σ_{cu}) is given by

$$\sigma_{cu} = \sigma_{fu} V_f \qquad (3.18)$$

Aveston et al.[10] have also shown from energy considerations that the matrix cracking strain (ϵ_{mu}) can theoretically be increased for fibre diameters below a critical value, at constant fibre volume.

3.4.2 Application to real composites

This section is based on work by Allen.[4]

The stress–strain curve shown in Figure 3.4 is for a matrix with a single valued cracking stress whereas, in reality, the matrix has considerable variability in strength from point to point. Also, the actual shape of the stress–strain curve will depend on the strength of the bond between the fibre and matrix and the volume of the fibres. These variables, together with the length efficiency factors and factors for fibre orientation will result in a range of stress–strain curves being formed in practice, all falling within the triangle OAB in Figure 3.4 where V_f has been modified by efficiency factors for length and orientation.[4,7] OB is the line for the relevant fibre volume alone with no contribution from the matrix.

Four very simplified curves for possible real composites are shown in Figure 3.6, where V'_f is the effective volume of fibre in the direction of stress calculated from the total V_f using appropriate efficiency factors. The values for (α) shown on Figure 3.6 have been calculated from equation (3.16) and, when substituted in equation (3.15), they give an indication of the possible range for real composites of the horizontal part ($\alpha\epsilon_{mu}/2$) of Figure 3.4

Figure 3.6(a) shows that it is possible for glass reinforced cement to have a typical extension due to multiple cracking alone of about 9 times the matrix cracking strain whereas asbestos cement (Figure 3.6(b)) can only extend by about 1 times the matrix cracking strain before the fibres take over completely. Likewise a similar volume of polypropylene to that of asbestos could increase the strain due to multiple cracking alone by nearly 50 times the matrix cracking strain even under fairly rapid loading. This would probably be exceeded for extended loading periods because the modulus of polypropylene is time dependent.

The other features of the stress–strain curves are that for glass reinforced cement (Figure 3.6(a)) the bond is fairly good, allowing many cracks to form (equation (3.14)), the matrix is fairly uniform but each matrix crack forms at a

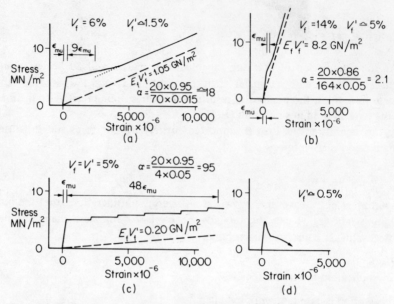

Figure 3.6. Theoretical stress—strain curves for four fibre types in cement based matrices (a) 2-D Random glass fibres in cement paste, (b) Asbestos fibres in cement paste, (c) Continuous aligned polypropylene fibres in cement paste, (d) Short, random, chopped steel or polypropylene fibres in concrete

slightly higher stress and the fibres are able to take over completely, shortly before fibre failure or pull out occurs.

For asbestos cement, (Figure 3.6(b)) the bond is very good resulting in many very fine cracks with only short debonded regions and a much smaller increase in total strain due to cracking as predicted by equation (3.15). Because the slope of the post-cracking stress strain curve is not greatly different from that of the matrix the transition between the two can be relatively smooth. Thus, the change in slope of the stress—strain curve is not always a good criterion by which the onset of multiple cracking may be judged. The lack of a horizontal portion in Figure 3.6(b) is a serious deficiency in that it results in the material having limited capability to absorb shocks and accidental over strains. It is also possible that the asbestos fibres have such a small diameter (0.02–20 μm) that matrix cracking is suppressed until successively higher strains are reached as suggested by Aveston,[10] and final failure is by single fracture with some fibre pull out before the zone of full multiple cracking is reached.

Figure 3.6(c) must necessarily have a very long horizontal portion according to equation (3.15) and every step may be long because of debonding at each crack. The slope $E_f V'_f$ may never be reached in practice due to excessive deformation.

Low volume, random short fibre, reinforced concretes are adequately represented by Figure 3.6(d) because when the matrix cracks, the fibres are unable to carry the cracking load and slowly pull out as the load is reduced.

3.4.3 Crack spacing and ultimate strength for short, aligned round fibres[1,2]

If, as before, it is assumed that after the matrix has cracked there is a linear transfer of stress from the fibres bridging the crack back into the matrix, then the theory in Section 3.4.1 can be modified to predict the properties of cement reinforced with short fibres. This section has been taken directly from the work of Aveston et al.[1,2]

For example the crack spacing may be calculated by considering the distance (x_d) required from the first crack for the fibres to transfer the load $(\sigma_{mu} V_m)$ per unit area of composite back into the matrix in the same way as the crack spacing for continuous fibres was calculated. The number of fibres with both ends at a distance greater than (x_d) from the crack and thus able to transfer their full share of the load is $N(1 - 2x_d/l)$ where $N = V_f/\pi r^2$. The remaining $(2x_d N)/l$ fibres have one end less than (x_d) from the crack and transfer load over a mean distance $(x_d)/2$. We put the total load transferred equal to $(\sigma_{mu} V_m)$ to give

$$2\pi r \tau x_d N \left(1 - \frac{x_d}{l}\right) = \sigma_{mu} V_m$$

or

$$x_d = \frac{1}{(1 - x_d/l)} \cdot \frac{V_m}{V_f} \cdot \frac{\sigma_{mu} r}{2\tau} = \frac{x'}{(1 - x_d/l)} \qquad (3.19)$$

where (x') is given by equation (3.14). Hence from equation (3.19)

$$x_d = l \pm \frac{(l^2 - 4lx')^{1/2}}{2} \qquad (3.20)$$

and as $x_d < l$ it is the smaller root that is required.

Equation (3.20) plotted in Figure 3.7 shows that the crack spacing is close to that of the continuously reinforced material until the fibre length approaches $(4x')$, i.e. until the mean transfer length $(l/4)$ equals the crack spacing for the continuously reinforced material.

The ultimate strength (σ_{cu}) will be less than $(\sigma_{fu} V_f)$ even when the fibre length (l) is greater than the shortest or critical length (l_c) given by

$$l_c = \frac{\sigma_{fu} r}{\tau} \qquad (3.21)$$

required to break the fibre in the matrix. This is because a proportion (l_c/l) of the fibres will have one end within a distance $(l_c/2)$ of a crack and will therefore pull out instead of breaking. The average stress in the fibres that pull out is equal to $(\tfrac{1}{2}\sigma_{fu})$ on the simple theory (section 3.4.4) and hence the ultimate tensile strength

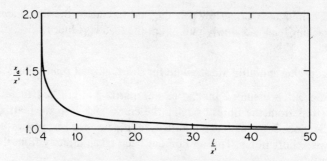

Figure 3.7. Crack spacing for short aligned fibres relative to the corresponding continuous case predicted by equation (3.20) (Reproduced from Aveston, Mercer, and Sillwood, *National Physical Laboratory Report No. SI*, No. 90/11/98 Part I, January 1975, by permission of the National Physical Laboratory. Crown copyright reserved)

will be reduced to

$$\sigma_{cu} = \left(1 - \frac{l_c}{2l}\right) \sigma_{fu} V_f \tag{3.22}$$

3.4.4 Short random round fibres which pull out rather than break

Factors affecting a realistic estimate of $V_{f(crit)}$ and post cracking strength are:

(a) Number of fibres across a crack and effective fibre orientation.
(b) Bond strength and fibre pull out load.

(a) Number of fibres across a crack and effective orientation.
The situation in a cracked composite may be represented by Figure 3.8.

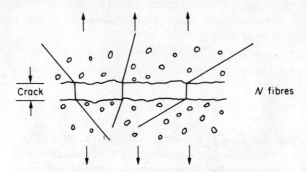

Figure 3.8. Change in fibre orientation at a crack

Just before cracking the efficiency factors (η_1) are as shown in Table 3.1 but after cracking and as the crack opens the fibres tend to pull into line with the stress[11,13] thus slightly increasing their efficiency as shown in Figure 3.8.

For short random fibres, such as steel or chopped fibrillated polypropylene, which are generally shorter than the critical length for fibre breakage, pull out mostly occurs across a crack. A realistic estimate of the load carried after cracking can therefore be obtained be multiplying the number of fibres crossing a unit area of crack by the average pull out force per fibre. For fibres which break before pulling out, the situation is more complicated. (see Sections 3.4.5 and 3.4.7).

The appropriate number of fibres ,'N' per unit area can be calculated as follows: [11] (Wire radius = r)

Aligned Fibres 1-D, $N = \dfrac{V_f}{\pi r^2}$ (3.23)

Random 2-D, $N = \dfrac{2}{\pi} \dfrac{V_f}{\pi r^2}$ (3.24)

Random 3-D, $N = \dfrac{1}{2} \dfrac{V_f}{\pi r^2}$ (3.25)

(b) Bond strength and fibre pull out force

If the composite failure is by fibre pull out it has been shown that the mean fibre pull out length is $(l/4)$[11], see Figure 3.9.

Provided that the average sliding friction bond strength (τ) is known and assuming that it does not vary with the angle of the fibre to the crack then the average pull out force per fibre (F) is given by:

$$F = \tau \pi d l / 4 \qquad (3.26)$$

The ultimate stress sustained by a unit area of composite after cracking is therefore given by N times F.

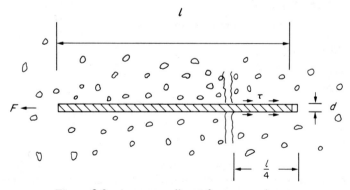

Figure 3.9. Average pull-out force per wire

i.e. $\sigma_{cu} = \dfrac{N\tau\pi dl}{4}$ (3.27)

and the average fibre stress at pull out (σ_f) is,

$$\sigma_f = \tau\pi d \dfrac{l}{4} \cdot \dfrac{4}{\pi d^2}$$

$$\sigma_f = \tau \cdot \dfrac{l}{d}$$ (3.28)

Substituting for N from equations (3.23), (3.24) and (3.25) in equation (3.27) gives

Aligned Fibres 1-D $\quad \sigma_{cu} = V_f \tau \dfrac{l}{d}$ (3.29)

Random $\quad$ 2-D $\quad \sigma_{cu} = \dfrac{2}{\pi} V_f \tau \dfrac{l}{d}$ (3.30)

Random $\quad$ 3-D $\quad \sigma_{cu} = \dfrac{1}{2} V_f \tau \dfrac{l}{d}$ (3.31)

Thus, the 3-D random fibre concrete should have about half the post crack strength of the aligned composite.

Equations (3.29), (3.30), and (3.31) can be used, either to deduce $V_{f(crit)}$ by equating to the right hand side of equation (3.11) or to deduce the ultimate strength of the composite for a known fibre volume fraction (V_f) and shear strength (τ).

3.4.5 Random discontinuous round fibres in a thin laminate

This case has been examined in detail by Allen[14] who has shown that the upper bound to strength is

$$\sigma_{cu} = \tfrac{1}{2}\eta_2 V_f \sigma_{fu}$$ (3.32)

where η_2 is defined by equations (3.8) and (3.9).

If we assume that

$$l_c = \dfrac{\sigma_{fu} \cdot d}{2\tau}$$

then for the case where $l \leqslant l_c$, equation (3.32) reduces to

$$\sigma_{cu} = \tfrac{1}{2} V_f \cdot \tau \cdot \dfrac{l}{d}$$ (3.33)

The predicted ultimate strength of a 2-D random composite with fibre pull out is therefore less than that given by equation (3.30) and the difference may be accounted for by different assumptions regarding the efficiency of fibres as they pull into line with the stress across a crack.

3.4.6 Long random round fibres

For the 2-D case Aveston et al.[11] have shown that the post-cracking modulus is

$$E_c = \frac{E_f V_f}{2} \qquad (3.34)$$

the ultimate strength is

$$\sigma_{cu} = \frac{\sigma_{fu} V_f}{2} \qquad (3.35)$$

and the equivalent crack spacing should be $\pi/2$ of the aligned model.

For the 3-D situation, the equivalent efficiency factor is $(2/3\pi)$.[12]

These efficiency factors only apply when the fibres are elastic and the composite breaks when the strain in the most highly strained fibres, i.e. those normal to the crack, reaches the fibre breaking strain.

3.4.7 Efficiency factors for a variety of fibre lengths and orientations

For composites in which the fibre length (l) exceeds the critical length (l_c) such as glass fibre reinforced cement, then a majority of the fibres may break before pulling out of the matrix. A detailed study of the appropriate efficiency factors (η_2) to be used in the post-cracking zone has been made by Laws[7] who has shown that the factor may vary between $(1 - l_c/2l)$ and $(1 - l_c/l)$ depending on the bond strength before fibre slip (τ_s) and after fibre slip (τ_d).

Total efficiency factors for post-cracking strength allowing for both fibre length and fibre orientation are tabulated below in Table 3.2 from the paper by Laws.[7]

These efficiency factors can be used with (V_f) in equations (3.2) and (3.5) to deduce composite properties after cracking if σ_m and E_m set to zero but their accuracy is limited, as usual, by the lack of accurate experimental data for τ_s and τ_d.

3.4.8 Realistic calculations for critical fibre volume ($V_{f(crit)}$) and ultimate composite strength (σ_{cu}) in direct tension

These examples use the same material parameters as in the idealized calculations in section 3.3.2.

Example 1. Realistic calculation of $V_{f(crit)}$ for short 3-D random steel fibres in concrete.

(a) Orientation of Fibres

Just before cracking, the efficiency factors are shown in Table 3.1 and using these values the actual $V_{f(crit)}$ for 3-D random fibres would be expected to be about, 5 × 0.31 per cent = 1.55 per cent, i.e. five times the value in Example 1, Section 3.3.2.

Table 3.2. Efficiency factors for post-cracking strength for restrained fibrous mat (Reproduced from Laws,[7] *Journal of Physics D: Applied Physics*, **4**, 1737–1746 (1971) by permission of the Institute of Physics)

Orientation	Efficiency factor		
	Continuous fibres	Short fibres	
Aligned	1	$\dfrac{l}{4l_c'}$	$(l \leqslant 2l_c')$
		$1 - \dfrac{l_c'}{l}$	$(l \geqslant 2l_c')$
Random 2-D	$\tfrac{3}{8}$	$\dfrac{9l}{80l_c'}$	$(l \leqslant \tfrac{5}{3}l_c')$
		$\tfrac{3}{8}\left(1 - \tfrac{5}{6}\dfrac{l_c'}{l}\right)$	$(l \geqslant \tfrac{5}{3}l_c')$
Random 3-D	$\tfrac{1}{5}$	$\dfrac{7l}{100l_c'}$	$\left(l \leqslant \dfrac{10l_c'}{7}\right)$
		$\tfrac{1}{5}\left(1 - \dfrac{5l_c'}{7l}\right)$	$\left(l > \dfrac{10l_c'}{7}\right)$

where $l_c' = \dfrac{l_c}{2}\left(2 - \dfrac{\tau_d}{\tau_s}\right)$

τ_s = static interfacial bond force.
τ_d = sliding frictional bond force.

Footnote: In a personal communication, Laws has stated that when $l \leqslant l_c$ a higher factor, $\beta \cdot (\tau_d/\tau_s) \cdot (l/l_c)$, where $\beta = \tfrac{1}{4}$ for 2-D or $\beta = \tfrac{1}{6}$ for 3-D fibres, can in theory apply in some cases if the requisite high strain ratio can be achieved.

However, this is still inaccurate because full bond and fibre breakage have been assumed, and also, if full bond is not achieved the orientation will change across the crack (Figure 3.8 above)

(b) Number of fibres across the crack and average pull out force.

The stress (σ_{cu}) in the composite carried by the fibres after cracking can be calculated from the average numbers of fibres across a unit area of crack multiplied by the fibre pull out force as in equation (3.31) i.e.

$$\sigma_{cu} = \tfrac{1}{2} V_f \tau \dfrac{l}{d} \tag{3.31}$$

Equation (3.31) is critically dependent on the average sliding frictional shear stress, (τ), at the fibre—cement interface and on the l/d ratio. The bond stress depends on a variety of factors such as w/c ratio, curing conditions, wire surface characteristics, wire geometry, and age. Measured values for τ have varied between 5.4 to 8.3 MN/m² [11], 3.9 MN/m² [12], 3.5 to 4 MN/m² [15] and 3 to 6 MN/m² [16] and for the purposes of this calculation a conservative estimate of τ of 3.5 MN/m² is taken.

As a first approximation for low fibre volumes the composite may be expected to crack at about the same stress as the matrix (Section 3.2)

i.e.

$$\sigma_{mc} \simeq \sigma_{cu} \simeq 3.0 \text{ MN/m}^2$$

(i) Let $\dfrac{l}{d} = 100$

Therefore

$$3 = \tfrac{1}{2}(V_{f\text{crit}}) \times 3.5 \times 100$$

Therefore

$$V_{f(\text{crit})} = 1.71 \text{ per cent}$$

(ii) Let $\dfrac{l}{d} = 50$

Therefore

$$V_{f(\text{crit})} = 3.42 \text{ per cent}$$

It should also be noticed that the stress (σ_f) in the wire at pull out is given by

$$\sigma_f = \tau \cdot \frac{l}{d} \qquad (3.28)$$

For cases (i) and (ii) above the stress in the wire at composite failure is only 350 MN/m² to 175 MN/m² respectively which may be less than one quarter of the steel strength thus resulting in a rather an inefficient use of the most expensive constituent in the composite.

Even these calculations are liable to considerable error because of other factors which are at present unquantifiable. These include the effect of multifibre pull out at random angles and their effect on matrix breakdown (References 17 and 18). The latter is particularly a problem for stiff crimped fibres where the fibre stress at pull out for single fibres is a high proportion of the ultimate strength but multifibre pull out can cause complete matrix breakdown at a relatively low fibre stress, probably due to the radial tensile stresses at the crimps.

However, the calculations demonstrate that the critical fibre volume is difficult to achieve in steel fibre concrete, because mixing and compaction problems increase considerably above about one per cent by volume, and also it is apparent that any theoretical assessment of the tensile strength of a real composite must, of necessity, be very imprecise.

Example 2: Realistic calculation of $V_{f(crit)}$ for glass reinforced cement and a description of post cracking behaviour

The correction factors for glass fibre cement are rather different from those for steel fibre concrete although the parameters affecting fibre efficiency are similar. Also two different approaches can be used to obtain an approximate estimate of the critical fibre volume.

(i) (a) Orientation Effects

The spray suction process for the manufacture of boards tends to give a 2-D random array and the efficiency factor of $\frac{3}{8}$ (Table 3.1) has been shown to be in reasonable agreement with the test data.[8] However, the fibre orientation in test boards has been shown not to be truly random and strengths are higher in samples cut from the longitudinal direction than the transverse direction.

(b) Bond Strength, Fibre Pullout Load, Fibre Length

In glass-reinforced cement a majority of the fibre strands break before pulling out of the matrix and frictional forces then maintain a high proportion of the strand fracture load.[9] (A strand may be about 0.65 mm wide by 0.11 mm thick and consists of 204, 12.5 μm diameter filaments)

The critical strand length (l_c) is defined as twice the embedded length at fibre failure and Laws[7] has suggested that an efficiency of $[1 - (2l_c/3l)]$ might be appropriate for glass-reinforced cement. The critical length for multifilament strand in cement will vary depending on age and storage conditions and figures of 12 mm and 26 mm have been published for l_c by Majumdar[8] and Oakley[9] respectively. These values depend on the bond strength assumed in the calculation. Chopped glass strand is commonly 34 mm long (l) and hence the efficiency may vary between 0.48 and 0.76 depending on the choice of (lc).

A very approximate estimate for the critical fibre volume in a real 2-D glass reinforced cement may therefore be calculated using the $V_{f(crit)} = 0.4$ per cent from Section 3.3.2 as follows:

$$V_{f(crit)} \simeq 0.4 \, \frac{1}{\frac{3}{8}} \, \frac{1}{(0.75 \text{ or } 0.48)}$$

$$\simeq \underline{1.4 \text{ per cent to } 2.2 \text{ per cent}}$$

(ii) An alternative method for calculating $V_{f(crit)}$ could be deduced from the work of Oakley and Proctor[9] for the particular case of spray de-watered pseudo random 2-D reinforcement with a sand/cement ratio <0.6.

These authors justified experimentally, the use of empirical efficiency factors to describe the overall effectiveness of random short fibres after the completion of multiple cracking and they have suggested the following equations:

In the longitudinal direction:

$$\text{Post-cracking modulus} \quad E_c \simeq 0.26 \, E_f V_f \qquad (3.36)$$

$$\text{Strength} \quad \sigma_c \simeq 0.27 \, \sigma_f V_f \qquad (3.37)$$

In the lateral direction

$$E_c \simeq 0.16 \, E_f V_f \qquad (3.38)$$

$$\sigma_c \simeq 0.17 \, \sigma_f V_f \qquad (3.39)$$

Substituting equation (3.37) in equation (3.11) and using the appropriate orientation factors on the R.H.S.

$$0.27 \, \sigma_f V_{f(crit)} = \epsilon_{mc} E_f \tfrac{3}{8} V_{f(crit)} + \sigma_{mc}(1 - \tfrac{3}{8} V_{f(crit)})$$

Therefore
$$V_{f(crit)} = \frac{\sigma_{mc}}{0.27 \, \sigma_f - \epsilon_{mc} \tfrac{3}{8} E_f + \tfrac{3}{8} \sigma_{mc}} \qquad (3.40)$$

Using realistic values for the various parameters

$$V_{f(crit)} = \frac{5 \times 10^6}{0.27 \times 1250 \times 10^6 - 300 \times 10^{-6} \times \tfrac{3}{8} \times 70 \times 10^9 + \tfrac{3}{8} \times 5 \times 10^6}$$

$$V_{f(crit)} \simeq \underline{1.5 \text{ per cent}}$$

This agrees well with the previous estimate

In contrast with steel-fibre concrete, 1.5 per cent of fibres can easily be included in glass-reinforced cement by the spray suction process and it would therefore be expected that this material could have an increase in load capacity in direct tensile stress fields after cracking and this has been confirmed in tests.

Example 3: Realistic calculation of $V_{f(crit)}$ for 3-D random, short chopped Fibrillated Polypropylene Fibre, in concrete.

Similar arguments apply as for examples 1 and 2 above although allowances for bond and critical length are not well documented. However, it is difficult to include more than one per cent by volume in concrete and compact the material using standard techniques so there is little point in guessing at a real critical fibre volume which will inevitably be greater than one per cent. Also, the strains and crack widths at which failure would be reached would probably be unacceptably large.

3.5 SPECIFIC FIBRE SURFACE IN RELATION TO CRACK SPACING

An empirical approach to crack spacing has been proposed by Krenchel[19] who has suggested that crack spacing may be linearly related to the specific surface of

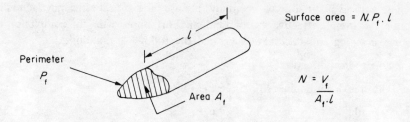

Figure 3.10. 'N' fibres in a unit volume of composite

the included fibres. This work was based on results produced for ferro-cement by Shah and Key[20] for specific surfaces between 0.1 and 0.5 mm²/mm³ and has been extended by Krenchel to fibre cements with specific fibre surfaces up to 5–10 mm²/mm³.

The specific fibre surface (S.F.S.) is defined as the total surface area of all fibres within a unit volume of composite (Figure 3.10). The end faces of the cut fibres are neglected and no account is taken of fibre orientation.

Total fibre surface in a unit volume of composite is therefore,

$$\frac{V_f}{A_f \cdot l} \cdot P_f \cdot l,$$

i.e. $\text{S.F.S.} = \dfrac{P_f}{A_f} \cdot V_f$ \hfill (3.41)

In the special case of cylindrical fibres we have

$$\text{S.F.S.} = \frac{\pi d}{\frac{\pi}{4} \cdot d^2} \cdot V_f = \frac{4}{d} V_f \tag{3.42}$$

Krenchel has suggested that the final average crack spacing (x'') is given by equation

$$x'' \times \text{S.F.S.} = \text{Constant} \sim 2.5 \tag{3.43}$$

It has been shown[19] that a typical glass reinforced cement may have a specific fibre surface of about 6.6 mm²/mm³ whereas asbestos cement may reach 50 mm²/mm³. Typical ordinary reinforced concrete may be between 0.005 and 0.020 mm²/mm³.

It is implied by equation (3.43) that crack spacing of 0.05 mm may be achieved for S.F.S. of 50 mm²/mm³ or 5.0 mm for S.F.S. = 0.5 mm²/mm³ which covers the range within which many fibre cement composites lie.

An advantage of a very high specific surface is that the matrix can be made to behave as a homogenous ductile material after matrix cracking has nominally occurred with apparently no discrete cracks in the tensile zone.

In fact, Krenchel's equation (3.43) is closely related to Aveston's equation (3.14).

Re-writing equation (3.43) for circular fibres

$$x'' \cdot \frac{4}{d} \cdot V_f = \text{constant}$$

Rewriting equation (3.14)

$$x' \cdot \frac{4}{d} \cdot V_f = \frac{\sigma_{mu} V_m}{\tau}$$

For many fibre composites (V_m) is between 0.9 and 1.0 and therefore, if Krenchel's equation is accurate, the implication is that σ_{mu}/τ is approximately constant for common fibre cements. Although this may be true for individual fibre types of different diameters in the same matrix it is unlikely to be a universal constant because poorly bonded organic fibres will probably have a ratio in excess of 2.5 whereas well bonded inorganic fibres may have σ_{mu}/τ closer to unity.

Another difference between the two approaches is that Aveston's can take account of fibre orientation, but this factor is not specifically included in the theory of Krenchel.

3.6 FRACTURE MECHANICS APPROACH TO FIBRE STRENGTHENING

An alternative approach to the theoretical prediction of fibre concrete cracking stress in direct tension was published by Romualdi and Batson[21] in 1963 based on the assumption that concrete is a notch-sensitive medium in which the critical flaw size can be calculated. A fracture arrest approach was adopted which indicated that, for a given volume of steel fibre, the direct tensile strength of the composite would increase with decreasing wire diameter and hence decreasing wire spacing. This type of material behaviour does not comply with normal reinforced concrete theory which predicts no change in the strength of the composite for a constant volume of steel. Aveston, Cooper, and Kelly[10] used a different approach to achieve a similar result[22] but it has been pointed out that both theories rely on the fibres and concrete being fully bonded together which leads to unrealistically high shear stresses at the interface with inevitable debonding.[12]

Thus neither theory is appropriate for the prediction of the tensile properties of fibre reinforced concrete.

An example of the lack of agreement between the fracture mechanics theory of Romualdi and experimental results is shown in Figure 3.11.

3.7 CALCULATION OF FIBRE SPACING

Since Romualdi[21] carried out his original work on the effects of fibre spacing on the properties of the hardened composite, there has been continuing controversy, not only regarding the effects of spacing on strength but also regarding the correct method of calculating the fibre spacing.

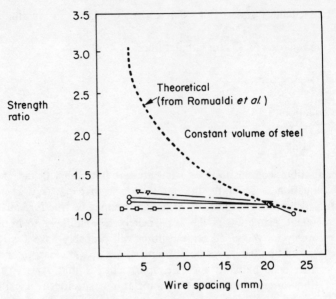

Figure 3.11. Effect of spacing of reinforcement on cracking strength of concrete
Experimental results
○—○ Shah and Rangan[23]
▽—▽ Johnston and Coleman[24]
□—□ Edgington[25]
(Reproduced from Edgington, Hannant, and Williams, *Building Research Establishment Current Paper CP 69/74*, July 1974, by permission of The Controller, HMSO. Crown copyright reserved)

The spacing may be calculated either from the distance between the centroids of indivdual fibres[15] or from the number of fibres crossing a unit area of a given plane section through the material[19] and the fibre array may be triangular, square, hexagonal, or other random patterns.

A logical argument for the plane section case is given by Krenchel[19] and the spacing equations given below are taken from Krenchel's paper for the particular case of cylindrical fibres in a square array.

1-D Parallel fibres. Plane section perpendicular to the direction of orientation

$$\text{Spacing} = \frac{\sqrt{\pi}}{2} \frac{d}{\sqrt{V_f}} \simeq 0.885 \frac{d}{\sqrt{V_f}} \tag{3.44}$$

2-D Random fibres. Plane section perpendicular to plane of orientation

$$\text{Spacing} = \frac{\pi}{2\sqrt{2}} \frac{d}{\sqrt{V_f}} \simeq 1.11 \frac{d}{\sqrt{V_f}} \tag{3.45}$$

3-D   Random fibres

$$\text{Spacing} = \sqrt{\frac{\pi}{2}} \frac{d}{\sqrt{V_f}} \simeq 1.25 \frac{d}{\sqrt{V_f}} \qquad (3.46)$$

REFERENCES

1. Kelly, A., *Strong Solids*, Clarenden Press, Oxford, 1973.
2. Holiday, L., *Composite Materials*, Elsevier, Amsterdam, 1966.
3. Holister, G. S., and Thomas, C., *Fibre-reinforced Materials*, Elsevier, Amsterdam, 1966.
4. Allen, H. G., 'Glass-fibre reinforced cement, strength and stiffness,' *CIRIA Report 55*, September, 1975.
5. Cox, H. L., 'The elasticity and strength of paper and other fibrous materials,' *British Journal of Applied Physics*, 3, 72–79 (1952).
6. Krenchel, H., *Fibre Reinforcement*, Akademisk Forlag, Copenhagen, 1964.
7. Laws, V., 'The efficiency of fibrous reinforcement of brittle matrices, *Journal Physics D: Applied Physics*, 4 1737–1746 (1971).
8. Majumdar, A. J., and Nurse, R. W., 'Glass-reinforced cement.' *Current Paper CP. 79/74*. Building Research Establishment.
9. Oakley, D. R., and Proctor, B. A., 'Tensile stress–strain behaviour of glass-fibre reinforced cement composites,' *Fibre-reinforced Cement and Concrete*, RILEM Symposium, 1975, Construction Press Ltd. London pp. 347–359.
10. Aveston, J., Cooper, G. A., and Kelly, A., 'Single and multiple fracture. The properties of fibre composites,' *Conference Proceedings of N.P.L. Conference*, IPC Science and Technology Press Ltd., 1971, pp. 15–24.
11. Aveston, J., Mercer, R. A., and Sillwood, J. M., 'Fibre-reinforced cements – scientific foundations for specifications Composites standards testing and design,' *National Physical Laboratory Conference Proceedings*, April 1974, pp. 93–103.
12. Aveston, J., Mercer, R. A., and Sillwood, J. M., 'The mechanism of fibre-reinforcement of cement and concrete.' *National Physical Laboratory Report No. SI*, No. 90/11/98, Part 1, January 1975; Part 11, DMA 228, February, 1976.
13. Argon, A. S., and Shack, W. J., 'Theories of fibre cement and fibre concrete.' *Fibre-reinforced Cement and Concrete*. RILEM Symposium, 1975, pp. 39–55.
14. Allen, H. G., 'The strength of thin composites of finite width, with brittle matrices and random discontinuous reinforcing fibres.' *Journal Physics D: Applied Physics*. 5, 331–343 (1972).
15. Swamy, R. N., Mangat, P. S., and Rao, S. V. K., 'The mechanics of fibre reinforcement of cement matrices.' *Fibre-reinforced Concrete*, American Concrete Institute Publication SP-44, pp. 1–28.
16. Hannant, D. J., 'Additional data on fibre corrosion in cracked beams and theoretical treatment of the effect of fibre corrosion on beam load capacity.' *Fibre-reinforced Cement and Concrete* RILEM Symposium, 1975, Volume 2, Construction Press, 1976, pp. 533–538.
17. Naaman, A. E., and Shah, S. P., Bond studies on oriented and aligned steel fibres.' *Fibre-reinforced Cement and Concrete*, RILEM Symposium, 1975, Construction Press Ltd. London pp. 171–178.
18. Hughes, B. P., and Fattuhi, N. I., 'Fibre bond strengths in cement and concrete.' *Magazine of Concrete Research*, 27 (**92**), pp. 161–170. (1975). September.

19. Krenchel, H., 'Fibre spacing and specific fibre surface.' *Fibre-reinforced Cement and Concrete*, RILEM Symposium, 1975. Construction Press Ltd., pp. 69–79. Volume 2, 1976, pp. 511–513.
20. Shah, S. P., and Key, W. H., 'Impact resistance of ferro-cement.' *Journal Structural Division., A.S.C.E.*, Vol. 98, No. ST1; *Proc. Paper. 8640*, January 1972, pp. 111–123.
21. Romualdi, J. P., and Batson, G. B., 'Mechanics of crack arrest in concrete,' *Proceedings A.S.C.E.*, Vol. 89, No. EM3, June 1963, pp. 147–168.
22. Kelly, A., 'Some scientific points concerning the mechanics of fibrous composites. *Composites–Standards, Testing, and Design*, National Physical Laboratory Conference, April 1974, pp. 9–16
23. Shah, S. P., and Rangan. B. V., 'Effects of reinforcement on ductility of concrete,' *Proceedings of the American Society of Civil Engineers: Journal of the Structural Division*, 1167–1184. June (1970).
24. Johnston, C. D., and Coleman, R. A., 'Strength and deformation of steel-fibre reinforced mortar in uniaxial tension', *American Concrete Institute Publication SP-44*, 1974, pp. 177–194.
25. Edgington, J. Hannant, D. J., and Williams, R. I. T., 'Steel-fibre reinforced concrete,' *Current Paper CP 69/74 Building Research Establishment*, July 1974.

Chapter 4
Theoretical Principles of Fibre Reinforcement in Flexure

4.1 GENERAL BACKGROUND

Many of the major applications of cement-bound fibre composites are likely to be subjected to flexural stresses in addition to direct stresses, and hence an understanding of the mechanism of strengthening in flexure may be more important than an analysis of the direct stress situation. In the following discussion the terms flexural strength, bending strength, and modulus of rupture are synonymous.

The need for a special theoretical treatment for flexure arises because of the large differences which are observed experimentally between the modulus of rupture and the direct tensile strengths, both in glass reinforced cement and in steel-fibre concrete. In both of these materials the so called 'modulus of rupture' can be up to three times the direct tensile strength even though, according elastic theory they are nominally a measure of the same value. The same situation occurs to a lesser degree with plain concrete.

The main reason for the discrepancy in fibre–cement composites is that the post-cracking stress–strain curve (Figure 3.6) on the tensile side of a fibre cement or fibre concrete beam is very different from that in compression and, as a result, conventional beam theory is inadequate. The flexural strengthening mechanism is mainly due to this quasi-plastic behaviour of fibre composites in tension as a result of fibre pullout or elastic extension of the fibres after matrix cracking, and the main principles are outlined in Figure 4.1.

Consider a fibre reinforced beam subjected to an increasing load (P) as shown in Figure 4.1. As the tensile strain increases, cracks are formed but, unlike plain cement or concrete, a proportion of the load is maintained across the crack by those fibres spanning the crack and hence equilibrium is maintained. Due to the formation of these cracks, the measured tensile strains increase and the value of (d_n), the distance of the neutral axis from the tensile surface, increases. As further load is applied to the beam, the measured tensile strains increase at a greater rate than the compressive strains (Figure 4.1(b)) until there is no simple relationship

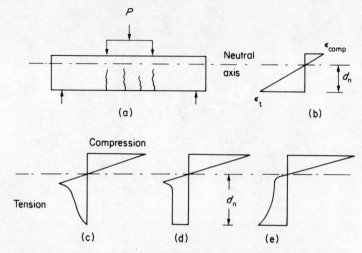

Figure 4.1. Strain and stress distributions in cracked fibre-cement or fibre-concrete sections subjected to flexure. (a) Flexural specimen under load, (b) Average strain distribution after cracking, (c) (d) (e) Alternative stress blocks depending on type of composite and volume fraction of fibres.

between the measured strain and the apparent stress sustained across the crack. The stress block in the tensile zone will then probably be approximated by one of Figures 4.1(c), (d), or (e), depending on the type of composite considered and whether the fibres carry a greater or lesser load across the crack than that sustained by the composite before cracking.

Figures 4.1(c) or (d) are more likely for steel fibre concrete, where the volume of fibre is probably less than the critical fibre volume in direct tension ($V_{f(crit)}$), whereas Figures 4.1(d) or (e) may be more appropriate for glass reinforced cement where the fibre content is just below or well above $V_{f(crit)}$ respectively. As a first approximation it is assumed that the compressive stress block is triangular although this may not be entirely accurate at ultimate load for fibre volumes above $V_{f(crit)}$.

However, even if the fibre volume for strengthening in tension has not been reached, it is still possible to increase the load that the beam will carry in flexure[1] due to the increased area of the tensile stress block caused by the pseudo-ductility and movement upwards of the neutral axis.

Thus, although values of modulus of rupture are often quoted for fibre cements and concretes based on elastic theory, these are not real values nor are they representative of tensile strengths. Even if the areas and shapes of the tensile stress blocks are accurately assessed the calculated tensile stresses are not real quantities. This is because the 'real' quantities are the forces in the individual fibres spanning cracks and these are effectively integrated, averaged and divided by the beam cross-sectional area to give a quantity which is convenient for engineering design, known as the average tensile stress in the composite. This is the same convenient quantity as is measured in a direct tensile test after matrix cracking and should not be confused with the modulus of rupture of an elastic material. The 'real' stresses in the fibres are likely to assume increasing significance as time, weathering, and other chemical actions reduce the fibre strength to near or below the fibre stresses and, when this occurs, failure of the composite may be initiated.

4.2 THEORETICAL ANALYSIS OF THE POST-CRACKING FLEXURAL BEHAVIOUR OF FIBRE CEMENT AND FIBRE CONCRETE

An accurate prediction from first principles of the flexural strength of fibre composites presents formidable problems because a series of assumptions of dubious accuracy has to be made. These include the effects of bond behaviour under strain gradients, the effects of fibre—fibre and fibre—aggregate interaction, fibre dispersion and orientation and length efficiency factors. However, an accurate theoretical treatment has been attempted by Aveston, Mercer, and Sillwood[2] using a direct tensile stress—strain curve predicted theoretically.[3] This theory[2] implies that the modulus of rupture may be up to three times the ultimate tensile strength for common fibre composites provided that failure is not initiated at the compression surface.

An approximate treatment which is similar in some of the principles to that of Aveston has been given by Hannant[1] and is used below to demonstrate the order of the improvements in moment of resistance of a section which can be achieved by utilising the post-cracking ductility provided by fibre pullout or fibre extension across a crack.

4.2.1 Analysis Using a Rectangular Stress Block in the Tensile Zone of a Beam

The analysis which follows is based on a simplified assumption regarding the shape of the stress block in the tensile zone after cracking. This assumption will inevitably be incorrect in varying degrees for every fibre cement composite as the shape will change with fibre type, fibre volume, fibre length, water/cement ratio, age, curing conditions, and crack width.

Nevertheless, the principles are considered to be worth stating, and in many cases of real composites, small changes in the shape of the stress block from that assumed will not greatly affect the conclusions drawn.

Figure 4.2(a) is for an elastic material with the neutral axis at mid-depth and tensile cracking strength of the composite (σ_t) equal to the modulus of rupture.(σ_{MR}). Figure 4.2(b) shows a stress block typical of a fibre concrete composite after cracking, where the fibres are extending or are pulling out at constant load across a crack throughout the tensile section. The ultimate post-cracking tensile strength of the composite (σ_{cu}) is the value calculated in Section 3.4 and (σ_{comp}) is the compressive stress on the outer face of the beam.

Figure 4.2(b) approximates to the stresses in steel-fibre concrete where the crack widths are small (<0.3 mm) compared with the fibre length and possibly to glass-reinforced cement at early ages when the fibres are poorly bonded and extend before fracture or pullout after fracture at roughly constant load[4].

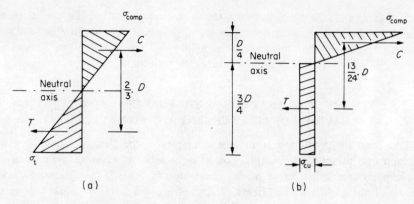

Figure 4.2. Stress blocks in flexure (a) Elastic material. Moment of resistance = $\sigma_t.D^2/6$ (b) Elastic in compression. Plastic in tension. Moment of resistance = $\sigma_{cu}\frac{13}{32}.D^2$ (Reproduced from Hannant,[1] *Fibre-reinforced Cement and Concrete, 2*, Figure 1 (1975) by permission of the publishers, The Construction Press Ltd.)

It has been shown by Edgington[5] for steel fibre concrete and by Allen[6] for glass-reinforced cement that the neutral axis may be only $0.2D$ from the compression surface a conservative estimate being $D/4$, and this assumption is used to make a quantitative assessment of several practical problems associated with the post cracking flexural strength of fibre concrete and fibre cement products.

In figure 4.2(a):

For equilibrium, forces $T = C$

$$T = \frac{\sigma_t}{2}\frac{D}{2} = \sigma_t\frac{D}{4}$$

Lever arm $l_a = \frac{2D}{3}$

Therefore

$$\text{moment of resistance} = \sigma_t\frac{D}{4}\frac{2D}{3}$$

$$= \frac{\sigma_t D^2}{6} \tag{4.1}$$

In figure 4.2(b):

Let (σ_{cu}) represent the force per unit area of section carried by the fibres, which is equivalent to the post cracking tensile strength in a direct tension test.

$$T = \sigma_{cu}\frac{3D}{4}$$

Lever arm $l_a = \frac{1}{2}\frac{3D}{4} + \frac{2}{3}\frac{D}{4} = \frac{13D}{24}$

Therefore

$$\text{moment of resistance} = \sigma_{cu} \frac{3D}{4} \frac{13D}{24}$$

$$= \sigma_{cu} \frac{13}{32} D^2 \qquad (4.2)$$

In order that the beams represented by Figures 4.2(a) and 4.2b) shall have the same strength, their moments of resistance should be equal i.e. from equations (4.1) and (4.2).

$$\frac{\sigma_t D^2}{6} = \sigma_{cu} \frac{13}{32} D^2$$

Therefore

$$\sigma_{cu} = \frac{16}{39} \sigma_t = 0.41 \sigma_t \qquad (4.3)$$

This implies that, provided the post cracking strength for large strains exceeds 0.41 of the tensile strength, then flexural strengthening can occur.

Thus, if the material has the tensile stress–strain characteristics shown in Figure 4.3(a), no decrease in flexural load capacity, or moment of resistance, will occur after cracking, and this is a common type of stress–strain curve for steel fibre concrete.

Provided that the distance (d_n) of the neutral axis from the tensile surface at failure is known, a similar approach can be used to deduce the appropriate (σ_{cu}), and hence the critical fibre volume for flexural strengthening for a range of (d_n). The critical fibre volume for flexure will always be less than or equal to ($V_{f(crit)}$) for direct tension.

A similar approach can be used to show why the modulus of rupture (σ_{MR}) calculated using an elastic analysis as in B.S. 1881[7] is generally found to be much

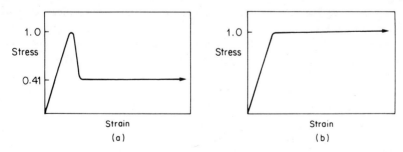

Figure 4.3. Stress–strain curves in uniaxial tension (a) No decrease in flexural load capacity after cracking, (b) Load capacity after cracking = 2.4 times cracking load for (compressive strength)/(tensile strength) ⩾ 6 (Reproduced from Hannant,[1] *Fibre-reinforced Cement and Concrete*, **2**, Figure 2 (1975) by permission of the publishers, The Construction Press Ltd.)

higher than the direct tensile strength. For instance, if the tensile stress strain curve in Figure 4.3(b) is assumed and the stress blocks in Figure 4.2(a) and 4.2(b) are for fibre beams made of the same material and failing at the same load, then the stress block in Figure 4.2(a) is an inaccurate representation of the real situation but is, nevertheless, generally used to calculate the modulus of rupture (σ_{MR}). Replacing (σ_t) in equation (4.3) by (σ_{MR})

$$\sigma_{MR} = \frac{39}{16} \sigma_{cu} = 2.44 \sigma_{cu} \tag{4.4}$$

Figure 4.3(b) is similar in form to some glass reinforced cements where the critical fibre volume for direct tension has just been achieved and the calculated modulus of rupture after cracking will be about 2.4 times the measured direct tensile strength. It follows that the beam would carry about 2.4 times the load which would be predicted from the use of the measured tensile strength in an elastic analysis for the load capacity of the beam.

It can be simply shown using a similar analysis that the maximum ratio (modulus of rupture)/(tensile strength) = 3. This occurs at the limiting condition when the neutral axis reaches the compression surface of the beam and the maximum post-cracking tensile strength of the composite (σ_{cu}) is maintained throughout the section depth, i.e.

$$\sigma_{cu} \cdot D \cdot \frac{D}{2} = \frac{\sigma_{MR} \cdot D^2}{6}$$

$$\frac{\sigma_{MR}}{\sigma_{cu}} = 3 \tag{4.5}$$

In practice this will rarely be achieved because compression failure will be initiated at the outer beam surface first.

An exact relation between the modulus of rupture and ultimate tensile strength of a brittle matrix composite has been given by Aveston, Mercer, and Sillwood[2] in terms of ($\alpha = E_m V_m / E_f V_f$) and the ratio of failure strain of the fibre to that of the matrix (ϵ_{fu})/(ϵ_{mu}). Figure 4.4 is reproduced from reference 2 and typical values of α are shown on Figure 3.6.

Equations 4.2, 4.3, and 4.4 are based on the assumption that adequate post cracking ductility is achieved (see section 4.5), and that failure is not initiated at the compression surface of the beam. The latter may be important when the neutral axis moves very close to the compression surface, but the compressive stress (σ_{comp}) can be calculated for Figure 4.2(b) as follows:

To prevent compressive failure: Equating $C = T$

$$\tfrac{1}{2} \sigma_{comp} \frac{D}{4} = \sigma_{cu} \frac{3D}{4}$$

$$\frac{\sigma_{comp}}{\sigma_{cu}} \geqslant 6 \tag{4.6}$$

This is usually the case for the matrix in fibre cement composites but it is

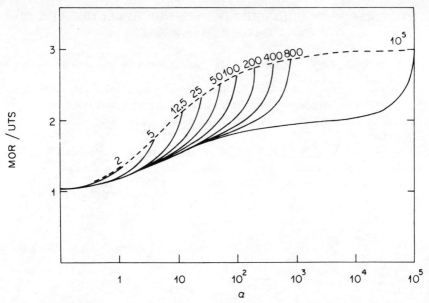

Figure 4.4. Ratio of the modulus of rupture to the tensile strength *vs.* $\alpha = E_m V_m / E_f V_f$ for various values of $\epsilon_{fu}/\epsilon_{mu}$ (Reproduced from Aveston, Mercer, and Sillwood, *Composites – Standards, Testing, and Design*, April 1974, pp. 93 – 103, by permission of the National Physical Laboratory. Crown copyright reserved)

possible for compression failure to occur at high fibre volumes because the fibres do not increase the compressive strength in proportion to the increase in tensile strength.

The basic theoretical treatment in Section 4.2 can be used to analyse a number of practical cases as in the following Sections 4.3 to 4.7.

4.3 MODULUS OF RUPTURE FOR ALIGNED FIBRE COMPOSITES

4.3.1 Continuous fibres

Using the assumptions given in Section 4.2.1, the post-cracking modulus of rupture (σ_{MR}) will be given approximately by substituting (σ_{cu}) from equation (3.18) into equation (4.4)

i.e. $\sigma_{MR} = 2.44\, \sigma_{fu} V_f$ \hfill (4.7)

4.3.2 Short fibres

As in Section 4.3.1, (σ_{cu}) from equation (3.22) is substituted in equation (4.4).

i.e. $\sigma_{MR} = 2.44 \left(1 - \dfrac{l_c}{2l}\right) \sigma_{fu} V_f$ \hfill (4.8)

4.4 FIBRES OF VARIOUS ORIENTATIONS WHICH PULL OUT AT A CRACK RATHER THAN BREAK[1]

4.4.1 Minimum Fibre Volume ($V_{f(min)}$) and l/d Ratio for Flexural Strengthening

Although factors other than flexural strength may control the performance of fibre concrete in bending, it may be important from economic considerations to be able to calculate the minimum fibre volume $V_{f(min)}$ necessary to achieve flexural strengthening.

Assumption: The fibres pull out of the matrix and the neutral axis is $3D/4$ from the tensile face after cracking.

Substituting equations (3.29), (3.30), and (3.31) for (σ_{cu}) in equation (4.3) gives:

1-D $\quad V_{f(min)} = \dfrac{16}{39} \dfrac{\sigma_t}{\tau} \dfrac{1}{(l/d)} \simeq 0.41 \dfrac{\sigma_t}{\tau} \dfrac{1}{(l/d)}$ (4.9)

2-D $\quad V_{f(min)} = \dfrac{8\pi}{39} \dfrac{\sigma_t}{\tau} \dfrac{1}{(l/d)} \simeq 0.64 \dfrac{\sigma_t}{\tau} \dfrac{1}{(l/d)}$ (4.10)

3-D $\quad V_{f(min)} = \dfrac{32}{39} \dfrac{\sigma_t}{\tau} \dfrac{1}{(l/d)} \simeq 0.82 \dfrac{\sigma_t}{\tau} \dfrac{1}{(l/d)}$ (4.11)

It has been shown in Sections 3.2 and 3.4.8 that (σ_t), the cracking stress in the composite, is approximately equal to (σ_{mu}) the cracking stress in the matrix for many practical systems and hence equations (4.9) to (4.11) imply:

(a) $V_{f(min)} \propto \sigma_t$ or σ_{mu} for a given (l/d), τ, and orientation.
(b) $V_{f(min)} \propto 1/(l/d)$ which is in general agreement with the experimental results of Edgington[8] for steel fibre concrete.
(c) $V_{f(min)}$ for 1-D fibre alignment is $\tfrac{1}{2} V_{f(min)}$ for 3-D alignment.

4.4.2 Apparent Post Crack Modulus of Rupture

The apparent modulus of rupture σ_{MR} after cracking can be calculated for a given fibre volume and shape by rearranging Equations (4.9), (4.10), and (4.11) and designating σ_t as σ_{MR} because the stress block is no longer elastic.

1-D $\quad \sigma_{MR} = V_f \tau \dfrac{l}{d} \dfrac{39}{16} \simeq 2.44 \, V_f \tau \dfrac{l}{d}$ (4.12)

2-D $\quad \sigma_{MR} = V_f \tau \dfrac{l}{d} \dfrac{39}{8\pi} \simeq 1.55 \, V_f \tau \dfrac{l}{d}$ (4.13)

3-D $\quad \sigma_{MR} = V_f \tau \dfrac{l}{d} \dfrac{39}{32} \simeq 1.22 \, V_f \tau \dfrac{l}{d}$ (4.14)

The significant implications from equations (4.12), (4.13), and (4.14) are as follows:

(a) The post-crack modulus of rupture is dependent only on the volume, shape and orientation of the fibres and the frictional bond strength and not on the strength of the matrix, although the latter may separately influence the bond strength and subsidiary cracking local to fibres.
(b) The apparent modulus of rupture increases in proportion to V_f and l/d. This agrees with the survey of a large amount of data by Johnston[9] for steel fibre concrete, although the precise relationship will depend on the fibre surface characteristics and test regime.
(c) The apparent modulus of rupture will be twice as great for 1-D fibre alignment as for 3-D fibre alignment. This is confirmed by the data of Hannant and Spring[10] for steel fibre concrete.

4.5 EFFECT OF LOSS OF DUCTILITY IN TENSION ON THE MODULUS OF RUPTURE

The strain to failure of some fibre cement composites, notably glass reinforced cement[11] may decrease drastically and continuously during a period of several years. This does not necessarily imply that the failure strain of the fibres has continuously decreased but rather that the bond between the fibre and the matrix has improved as hydration products are deposited to increase the contact area and frictional forces at the interface.[12] The strain in the composite at a given stress depends on the length of debonded fibre and hence a greater bond leads to a smaller failure strain in the composite.

This increase in bond can affect the composite strength in several ways:

(a) A well-bonded fibre crossing a matrix crack will be more highly stressed than a fibre with low bond because the strain at a crack is infinite and the crack opening will be transferred over a shorter debonded fibre length. This will result in less redistribution of stress across a section in a variable strength matrix with random cracks even under direct tension.
(b) Fibres at random angles across matrix cracks will suffer higher bending stresses as the cracks open due to increased strength and density of the matrix at the bending point of the fibre.[12] (Figure 3.8)
(c) The reduction in composite tensile failure strain can lead to a reduction in flexural strength even if the direct tensile strength remains the same.[1] The main principles of this approach are illustrated in Figures 4.5 and 4.6.

The important point to realize is that in order to achieve the stress block in Figure 4.2(b), the strain distribution in flexure must approximate to that in Figure 4.5(a) with a strain of about 18 ϵ_x in tension. The material represented by curve (a) on Figure 4.5 will fulfil this requirement whilst maintaining the maximum tensile load in the fibres at the outer beam surface, and thence the moment of resistance of the section = 0.41 $\sigma_{cu}D^2$. (equation (4.2) and Figure 4.6(a))

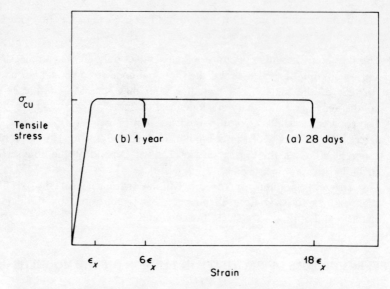

Figure 4.5. Idealized direct tensile stress–stress curve for fibre cement. ϵ_x is the strain at the start of major multiple cracking at the bend over of the stress–strain curve to near horizontal.

However, the strain capacity of the material represented by curve (b) in Figure 4.5 is insufficient to permit the section to reach the condition of the fully ductile stress block shown in Figure 4.2(b) because the outer fibres will start to break or pull out at a strain of 6 ϵ_x. The stress block will then be represented by Figure 4.6(b) and the load capacity will decrease thereafter because the outer tensile material with the largest lever arm will no longer carry load.

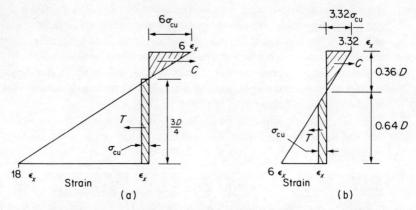

Figure 4.6. Strain and stress distributions in flexure for the tensile stress–strain curves in Figure 4.5 (a) For curve (a) in Figure 4.5, moment of resistance = 0.41 $\sigma_{cu}D^2$ (b) For curve (b) in Figure 4.5, moment of resistance = 0.35 $\sigma_{cu}D^2$

The moment of resistance for Figure 4.6(b) is approximately $0.35\,\sigma_{cu}D^2$ and since values of σ_{cu} are equal for both composites, this represents a reduction in load capacity of 15 per cent.

Thus it can be shown that an increase in fibre–cement bond, leading to a reduction in tensile strain capacity could, in some circumstances, cause a reduction in the ultimate modulus of rupture of the composite. This would occur if the strain capacity is insufficient to allow an adequate ductile stress block to develop.

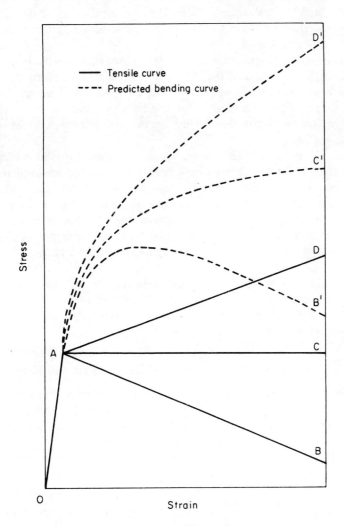

Figure 4.7. Apparent bending (modulus of rupture) curves predicted for assumed direct tensile curves (Reproduced from Laws and Ali,[13] *Conference on Fibre-reinforced Materials, 1977*, pp. 101–109, by permission of The Institution of Civil Engineers)

A calculation for a real composite is given in section 4.8.3 but the accuracy of such calculations is limited by a lack of knowledge of the appropriate stress–strain curve to use. For instance, measured stress–strain curves are generally obtained from transducers with gauge lengths between 10 mm and 100 mm and thus represent average strains in the composite. It may be more appropriate for the purposes of calculating the local position of the neutral axis to know the average strain over a gauge length of possibly less than 2 mm across a crack and this is likely to greatly exceed the strains shown in published stress–strain curves for the material, particularly for well-bonded fibres. Thus published stress–strain curves for well-bonded materials are likely to lead to an underestimate of the modulus of rupture if this approach is used.

A more comprehensive treatment showing the general effects of the shape of the tensile stress–strain curve and of the tensile strain capacity on the load capacity of fibre cement beams has been carried out by Laws and Ali[13] and is summarized in Figure 4.7. In this figure the 'stress' for the bending curves is effectively the modulus of rupture calculated from the bending moment assuming a linearly elastic beam in pure bending.

In Figure 4.7, the line OAB' represents the predicted apparent modulus of rupture–strain curve for the relatively poor type of tensile stress–strain curve OAB. This indicates that the maximum load point in bending will occur after the maximum tensile stress has been reached on the tensile beam surface provided that the tensile strain capacity is adequate. Also, the apparent modulus of rupture OAC' predicted from the tensile curve OAC continues to rise as the material yields until the ratio (Modulus of rupture)/(Tensile strength) approaches a value of 3 at large strains.

It is important to note that if the tensile strain capacity reduces with time and eventually starts to approach point A on Figure 4.7, then the implication is that the modulus of rupture may reduce very rapidly if the material looses its remaining quasi-ductility and becomes effectively elastic.

4.6 EFFECT OF SECTION DEPTH ON MODULUS OF RUPTURE

Using similar principles to those in Section 4.5 it is suggested that the modulus of rupture for deep sections is likely to be less than for thin sections. This is because, for the neutral axis to be at $3D/4$ from the tensile face, the primary cracks will be wider for deep sections than for thin sections. Thus, for the same material, the strain capacity may be exceeded locally before the neutral axis reaches $3D/4$ leading to a lower modulus of rupture.

4.7 EFFECT OF WIRE SPACING ON THE POST-CRACKING STRENGTH OF STEEL FIBRE CONCRETE

It is difficult to conceive of a fundamental reason why fibre spacing in its own right should have a unique relationship to post-cracking flexural strength because the matrix is carrying no stress across the crack.

Nevertheless, strength-spacing relationships are a matter for continuing controversy in studies of fibre concrete and equations (4.12), (4.13), and (4.14) have been utilized, in conjunction with the spacing equations of Krenchel[14] (given in section 3.7) to assess the theoretical relationship between post-crack modulus of rupture and fibre spacing.[1]

It was concluded from this theoretical work on fibre spacing-strength relationships that they are neither a useful concept, nor are they likely to further the understanding of post-cracking material behaviour because other factors, such as fibre length and bonded area, and fibre orientation, have much more fundamental significance than spacing.

4.8 APPLICATION OF FLEXURAL THEORY TO PRACTICAL COMPOSITES

4.8.1 Minimum fibre volume ($V_{f(min)}$) for flexural strengthening of steel fibre concrete

Equations (4.9) to (4.11) indicate that $V_{f(min)}$ is dependent on σ_t/τ i.e. the ratio of the tensile cracking stress in the composite (σ_t) to the sliding friction bond strength (τ).

In practice, σ_t is of the same order as τ and Figure 4.8 has therefore been plotted for $\sigma_t/\tau = 1$ for various fibre orientations. It can be seen from Figure 4.8 that flexural fibre strengthening can occur for practical composites between fibre volumes of about 0.3–1.3%, depending on the l/d ratio and orientation.

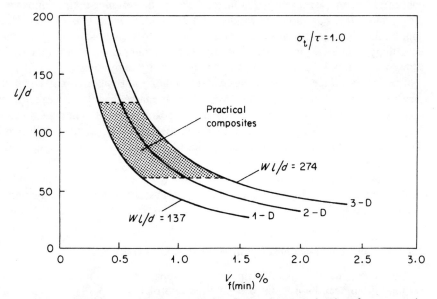

Figure 4.8. Minimum fibre volume for flexural strengthening for $\sigma_{t/\tau} = 1$ (Reproduced from Hannant,[1] *Fibre-reinforced Cement and Concrete*, 2, Figure 4 (1975) by permission of the publishers, The Construction Press Ltd.)

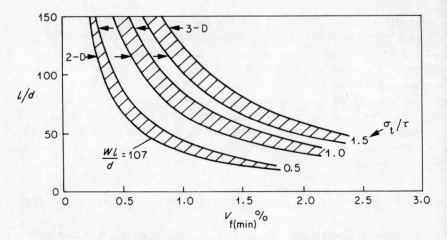

Figure 4.9. Minimum fibre volume for flexural strengthening for various σ_t/τ (Reproduced from Hannant,[1] *Fibre-reinforced Cement and Concrete*, 2, Figure 4 (1975) by permission of the publishers, The Construction Press Ltd.)

An interesting point about equations (4.9) to (4.11) and hence the curves shown in Figure 4.8 is that they imply that for a given σ_t and τ, the product of the l/d ratio and the minimum fibre volume for flexural strengthening will be constant for a particular fibre orientation. This finding tends to confirm the conclusions of Johnston[9] from a survey of experimental data that the product of fibre concentration and aspect ratio is a major factor in determining the performance of the composite. For normal weight concretes, the weight of fibre (W) is approximately proportional to the volume and, because the weight may be of more immediate interest to practising engineers than the volume, the parameter (Wl/d) is shown on Figures 4.8 and 4.9 where W = percent of fibres by weight.

The effect of reducing or increasing the cracking strength of the composite relative to τ is shown in Figure 4.9 from which it can be seen that a weak material such as wet lean concrete ($\sigma_t \simeq 1.0$ MN/m^2) can be reinforced in flexure at volume fractions less than 0.5% for a fibre l/d = 100. Alternatively, where the composite has a high cracking strength relative to the fibre bond strength, $V_{f(min)}$ can be considerably increased.

The parameter (Wl/d) is again shown on Figure 4.9, and it has been shown experimentally[9] that the minimum value of (Wl/d) at which fibre strengthening can occur is about 100.

4.8.2 Post-cracking modulus of rupture for steel fibre concrete

The shape of the relationship between apparent modulus of rupture and volume of wire can be calculated from equations (4.12) to (4.14) and is shown in Figure 4.10 for an assumed value of (τ) of 3.5 MN/m^2 (see Section 3.4.8). The form of Figure 4.10 is in agreement with extensive experimental work.[5,8,15]

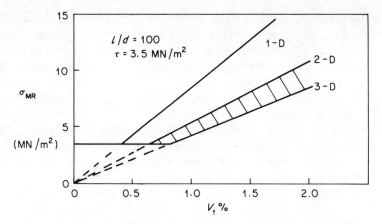

Figure 4.10. Theoretical apparent modulus of rupture — Effect of fibre orientation (Reproduced from Hannant,[1] *Fibre-reinforced Cement and Concrete*, 2, Figure 5 (1975) by permission of the publishers, The Construction Press Ltd.)

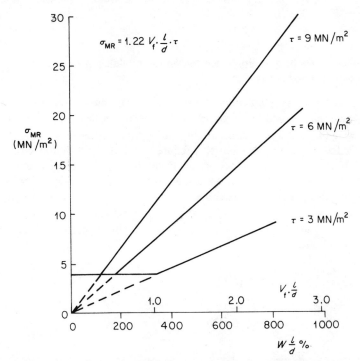

Figure 4.11. Effect of different sliding friction bond strengths. τ, on the theoretical post-cracking modulus of rupture for 3-D fibres (Assume matrix $\sigma_{MR} = 4$ MN/m^2)

A controlling factor in the calculation of the modulus of rupture from equations (4.12), (4.13), and (4.14) is the value of τ, the average sliding friction bond stress. A fibre shape which can increase τ by mechanical interlock with the paste, whilst maintaining some slip to give ductility and simultaneously being flexible enough to avoid breaking the matrix laterally is an ideal which is difficult to achieve in practice. The effect of varying the value of τ on the theoretical σ_{MR} is shown in Figure 4.11 for steel fibre concrete.

It is apparent from Figure 4.11 that a considerable increase in performance of the composite in bending could possibly be achieved by improving the fibre shape, i.e. increasing τ and several manufacturers have attempted to improve the bond of the wire by various mechanical treatments.

Although equations (4.12), (4.13), and (4.14) are independent of beam depth and length, the apparent moduli of rupture are likely to decrease in practice for beam depths greater than 100 mm[16]. This is probably because, for deep beams with the neutral axis at $D/4$ from the compression face, the primary cracks at the tensile surface will be wider than for shallow beams. The fibre pull out load will be decreased at the wider cracks and hence the ductile stress block will not be as efficient as that shown in Figure 4.2(b) and will lead to a lower apparent modulus of rupture. Increase in simply supported beam length can also reduce the modulus of rupture[16] Figure 6.5 and this may again be associated with the development of wider primary cracks.

4.8.3 Effect of loss of ductility in tension on the modulus of rupture of glass-reinforced cement

Figure (4.5) is a simplification of the stress–strain curves for glass-reinforced cement which may, in addition, exhibit a reduction in tensile strength with time.

Stresses and strains appropriate to glass-reinforced cement[17] are $\epsilon_x \simeq 500 \times 10^{-6}$, with composite failure strains of $9{,}000 \times 10^{-6}$ and $3{,}000 \times 10^{-6}$ at 28 days and 1 year respectively. If, σ_{cu}, the tensile strength of the composite is taken as 10 MN/m^2 then the moments of resistance in Figure 4.6(a) and (b) respectively will be $4.1D^2$ and $3.5D^2$.

The moment of resistance for an elastic material is $(\sigma_{MR} \cdot D^2)/6$ (see Figure 4.2a) and therefore the moduli of rupture of the two ages of G.R.C. can be predicted as follows:

28 days $\quad \dfrac{\sigma_{MR} \cdot D^2}{6} = 4.1D^2$

$\therefore \quad \sigma_{MR} = 24.6 \text{ MN/m}^2$

1 year $\quad \dfrac{\sigma_{MR} \cdot D^2}{6} = 3.5D^2$

$\therefore \quad \sigma_{MR} = 21.0 \text{ MN/m}^2$

This represents a reduction of 15% in modulus of rupture which is rather less than some reported values although these are also associated with a reduction in tensile strength.[11,18]

REFERENCES

1. Hannant, D. J., 'The effect of post cracking ductility on the flexural strength of fibre cement and fibre concrete', *Fibre-reinforced Cement and Concrete*, RILEM Symposium, Volume 2, Construction Press Ltd., 1975, pp. 499–508.
2. Aveston, J., Mercer, R. A. and Sillwood, J. M., 'Fibre-reinforced cements – scientific foundations for specifications,' *Composites – Standards, Testing and Design*, National Physical Laboratory Conference Proceedings, April 1974, pp. 93–103.
3. Aveston, J., Cooper, G. A., and Kelly, A., 'The properties of fibre composites,' *Conference Proceedings*, National Physical Laboratory, 1971, Paper 2, p. 15, IPC Science and Technology Press.
4. Oakley, D. R., and Proctor, B. A., 'Tensile stress–strain behaviour of glass fibre-reinforced cement composites.' *Fibre-reinforced Cement and Concrete*, RILEM Symposium, 1975, Construction Press Ltd., pp. 347–359.
5. Edgington, J., 'Steel-fibre reinforced concrete,' *PhD Thesis*, University of Surrey, 1973.
6. Allen, H. G., 'Stiffness and strength of two glass fibre reinforced cement laminates,' *Journal Composite Materials*, 5, 194–207 (1971), April.
7. British Standards 1881 (1970), *Part 4: Methods of Testing Hardened Concrete*.
8. Edgington, J., Hannant, D. J., and Williams, R. I. T. W., 'Steel fibre-reinforced concrete,' *Current Paper CP 69/74*, Building Research Establishment, July 1974.
9. Johnston, C. D., 'Steel-fibre reinforced mortar and concrete. A review of mechanical properties,' *Fibre-reinforced Concrete*, American Concrete Institute Publication SP-44, pp. 127–142.
10. Hannant, D. J., and Spring, N., 'Steel-fibre-reinforced mortar: a technique for producing composites with uniaxial fibre alignment,' *Magazine of Concrete Research*, 26(86), 47–48 (1974) March.
11. Building Research Establishment, A study of the properties of Cem-FIL/OPC composites. *Current Paper CP 38/76*, June 1976. 14 pp.
12. Stucke, M. S., and Majumdar, A. J., 'Micro-structure of glass fibre-reinforced cement composites,' *Journal of Materials Science*, 11, 1019–1030 (1976).
13. Laws, V., and Ali, M. A., 'The tensile stress/strain curve of brittle matrices reinforced with glass-fibre,' *Fibre-reinforced Materials Conference*, Insititution of Civil Engineers, London 1977, pp. 101–109.
14. Krenchel, H., 'Fibre spacing and specific fibre surface.' *Fibre-reinforced Cement and Concrete*, RILEM Symposium, 1975, Construction Press Ltd., pp. 69–80.
15. Moens, J., 'Steel wire fibre optimization. Fibre-reinforced concrete.' *Conference Proceedings*, Delft, September 1973, pp. 35–42.
16. Swamy, R. N., and Stavrides, H., 'Some properties of high workability steel fibre concrete,' *Fibre-reinforced Cement and Concrete*, RILEM Symposium, 1975, Construction Press Ltd., pp. 197–208.
17. Majumdar, A. J., 'Properties of fibre–cement composites,' *Fibre-reinforced Cement and Concrete*, RILEM Symposium, 1975, pp. 279–314.
18. Majumdar, A. J., and Nurse, R. W., 'Glass-fibre-reinforced cement,' *Current Paper CP 79/74*, Building Research Establishment, August 1974.

Chapter 5
Steel-Fibre Concrete: Properties in the Fresh State and Mix Design for Workability

5.1 GENERAL

In the developing field of steel-fibre concrete, the situation may arise in which an engineer wishes to make a quick assessment of the possible merits of steel fibres in a particular product or application. There is a great temptation in these circumstances to add fibres to an existing mix and to try to compare the new product with the existing one. Unfortunately, if the mix contains a normal proportion of aggregates of greater than 10 mm maximum size, both mixing and compaction problems generally result if a reasonable quantity of fibres is added, (about 1 per cent to 2 per cent by volume or 3 per cent to 7 per cent by weight) and the fibre concrete may be discarded as too difficult to produce. This practical experience has lead to the development of mix designs which will easily accept sufficient fibres of an appropriate type, which will give acceptable compaction characteristics and which can also provide useful properties in the hardened state.

5.2 ASSESSMENT OF WORKABILITY

The three commonly used tests for assessing the workability of plain concrete, i.e. the slump, compacting factor, and V–B tests[1] have all been used to assess the workability of fibre concrete.

5.2.1 Slump test

The slump test is only likely to be of value as a control test for fibre concretes for mixes in which the slump of the plain concrete exceeds 100 mm. This can apply to some of the American type mixes where P.F.A. (pulverized fuel ash) and air entraining agents have been used, or to mixes with super plasticizers. High strength mixes of the type often used in the United Kingdom tend to have nearly zero slump after the addition of fibres even although they may respond satisfactorily to vibration and hence the test is not of great value for this type of material.

5.2.2 Compacting factor

The compacting factor test has been tried for fibre reinforced mortars with only limited success[2] because the addition of fibres tends to prevent the material falling freely through the hoppers unless considerable force is applied, thus invalidating the concept of the test.

5.2.3 V–B consistometer

The V–B test, in which a slump cone of concrete is remoulded into a cylinder by vibration, is probably the best test to apply for stiff fibre concrete mixes because it simulates, at least in some respects, the compaction of concrete by vibration in practice. Also it identifies quite positively, a critical fibre volume for a given type of wire above which compaction is unlikely to be possible using normal site compaction procedures.

However, as for plain concrete, the most reliable assessment of workability for a particular application is to carry out a trial with the compaction plant to be used on the job.

5.3 EFFECT OF FIBRE AND AGGREGATE PARAMETERS ON WORKABILITY

5.3.1 Fibre length and diameter

A collection of long thin fibres of length/diameter greater than 100 will, if shaken together, tend to interlock in some fashion to form a mat, or a type of bird's nest from which it is very difficult to dislodge them by vibration alone. Short stubby fibres on the other hand of length/diameter less than 50 are not able to interlock and can easily be dispersed by vibration.

Similar effects are observed when fibres are dispersed in mortar or concrete and the ease with which the fibres can move relative to each other under vibration is shown in Figure 5.1 for mortars, with a particle size less than 5 mm.

It can be seen from Figure 5.1 that the l/d ratio has a crucial influence on the volume of fibres which can be included in the mix with relatively easy compaction (say V–B $<$ 20 seconds). It is shown in Section 3.4.8 that the critical fibre volume for strengthening in direct tension may be about 1.7 per cent at an l/d of 100 and Figure 5.1 indicates that practically this may only just be achieved with mortars, let alone with concretes. However, the fibre volume required for strengthening in flexure (Figures 4.8 and 4.9) can be achieved much more easily with acceptable compaction characteristics.

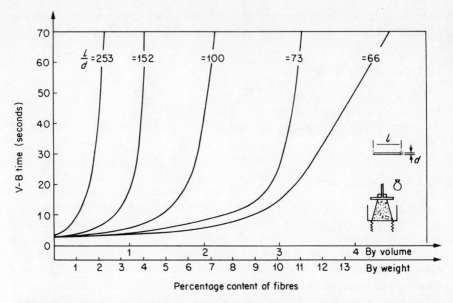

Figure 5.1. Effect of fibre aspect ratio on V–B time of fibre-reinforced mortar (Reproduced from Edgington, Hannant, and Williams, *Building Research Establishment Current Paper CP 69/74*, July 1974, by permission of The Controller, HMSO. Crown copyright reserved)

5.3.2 Aggregate size and volume

The problem is more complicated when fibres are introduced into a concrete rather than a mortar matrix because they are separated not by a fine grained material which can move easily between them, but by particles which will often be of a larger size than the average fibre spacing if the fibres were uniformly distributed. This leads to bunching and greater interaction of fibres between the large aggregate particles and the effect becomes more pronounced as the volume and maximum size of the particles increases. The principle is demonstrated in Figure 5.2.

Figure 5.2 shows diagramatically that uniform fibre dispersion is more difficult to achieve as the aggregate size increases from 5 mm to 10 mm to 20 mm. However, this is a simplified picture because, in reality, the fibre and aggregate dispersion is three dimensional and there may be up to 200 fibres in any given cube of mortar of side length equal to the fibre length before fibre interaction becomes excessive.[3] Nevertheless, it is apparent from Figure 5.2 that the greater the volume and size of the coarse aggregate, the more fibre interaction will occur.

In a normal concrete mix the particles finer than 5 mm occupy about 54 per cent of the volume, the 10 mm aggregate about 20 per cent and the 20 mm aggregate about 25 per cent of the volume. Thus only about 54 per cent of the real volume (i.e. the mortar fraction) is available for free fibre movement during

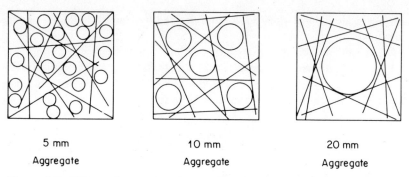

5 mm	10 mm	20 mm
Aggregate	Aggregate	Aggregate

Figure 5.2. Effect of aggregate size on fibre distribution within a square of side length = fibre length (40 mm)

compaction. Experience has shown that a satisfactory mix for fibre concrete should contain a mortar volume of about 70 per cent with only about 30 per cent consisting of particles between 5 mm and 10 mm.

The effect of a range of aggregate sizes and volumes on the compaction times of composites made with wires of $l/d = 100$ is shown on Figure 5.3.

Figure 5.3 indicates that for a V–B time of 20 seconds the 10 mm concrete will only accept about 50 per cent of the fibre volume compared with mortar and the 20 mm concrete will carry less than the 10 mm concrete. These particular results are due to a combination of the effects of aggregate size and aggregate volume because the mortar fraction is lowest for the 20 mm mix. However, the effect of

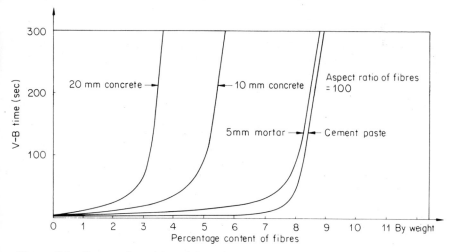

Figure 5.3. Compaction time against fibre content for matrices with different maximum aggregate size (Reproduced from Edgington, Hannant and Williams, *Building Research Establishment Current Paper CP 69/74*, July 1974, by permission of The Controller, HMSO. Crown copyright reserved)

coarse aggregate volume alone has been examined by Mangat[4] who has obtained a similar trend.

Thus it can be seen that the fibre sizes and volumes calculated in Chapters 3 and 4 for adequate hardened properties (i.e. high l/d ratio and large fibre volume) have to be carefully balanced against the mix design for adequate compaction properties which require a low l/d ratio and low fibre volume.

5.4 MIX DESIGN FOR WORKABILITY

5.4.1 Proportions of basic concrete mix

British practice has tended to avoid the use of additives to improve workability and has generally relied on a high sand content (more than 50 per cent by weight of the aggregate) with a maximum aggregate size of 10 mm. Typical mixes which have been used are shown in Table 5.1 (mixes 1 to 6) although Swamy and Stavrides[5] have experimented with mixes containing P.F.A. and water reducing admixtures. (Mix 7, Table 5.1)

American and Canadian practice, on the other hand, has been to use P.F.A. (Pulverized Fuel Ash) to increase the fines content and also to use air entraining agents and water reducing admixtures in addition to a sand content of 50 per cent or more. Some typical American mixes are shown in Table 5.1 (Mixes 8 to 11) and slumps of up to 150 mm can be achieved with these mixes at reasonable fibre volumes.

The introduction of superplasticizers has lead to trials to determine whether a higher volume of fibres can be included in a given mix by the use of these materials. It has been found that, although compaction is easier for a given fibre volume or alternatively more fibres can be included for a practical compaction time (say V–B $<$ 20 seconds), the maximum fibre volume which the mix will carry is not greatly altered. This is because the cement paste becomes more fluid with the addition of superplasticizers and tends to run out of the fibre clusters as they start to form. Segregation or clumping of fibres therefore occurs at about the same fibre volume as for the unplasticised matrix.

Thus, superplasticizers should be considered mainly as an aid to increase the workability of one of the standard types of mixes, or else to reduce the water/cement ratio of a high workability mix in order to achieve adequate strength and durability in the hardened state.

Whatever the basic mix design, trial mixes are essential to determine the workability and strength properties.

5.4.2 Fibre quantity

The factors affecting the maximum quantity of a particular fibre which can be included in a mix whilst maintaining adequate workability for site compaction have already been discussed in Section 5.3.

A simple empirical equation is given below which enables an approximate

Mix no.	Reference	Ratios by weight					Additives	Fibres		
		Water		Aggregate				Cross-section by length (mm)	Percentage by weight
		Free	Total	<5 mm	10 mm					
1	Edgington et al.[2]	0.40 0.40	0.43 0.48	2.4 2.04	— 1.36		—	Various		
2	National Standard[6]	0.40[a] 0.45[a]	— —	2.4 2.36	— 2.36		—	Various		
3	Johnson Nephew[7]	≯0.55[a] ≯0.55[a]	— —	2.4 2.0	— 2.0		—	Various		
4	Wakefield Slab. Mix C[8]	0.31[c]	—	2.5	0.82		6 per cent air	0.25 dia x 25	5.5
5	BOAC Car park[9]	0.65[a]	—	3.0	1.5		—	0.25 dia x 25	3
6	M.10 Overlay[10]	0.55[a]	—	2.15	2.15		—	0.5 dia x 38	up to 2.7
7	Swamy et al.[5]	—	0.6	2.6	3.2		0.43 P.F.A. Water reducing. Slump 90 mm	0.5 dia x 38	3.3
8	A.C.I. Mix[11]	0.54[a]	—	3.0	3.0		0.46 P.F.A. air. Water reducing. Slump 150 mm	0.25 x 0.56 x 25	5
9	[b]C.S.A.[12] Slab Overlay	0.50[a] 0.46[a]	— —	2.48 2.52	0.83 0.84		5.5 per cent air. Slump 125 mm 5.9 per cent air. Slump 80 mm	0.4 dia x 25 0.25 dia x 0.4 x 25	6.6 6.6
10	Tampa Airport[13]	0.53[a]	—	2.95	2.3 (20 mm)		0.44 P.F.A. Air. Water reducing	0.25 x 0.5 x 25	5.1
11	Calgary Slabs[14]	—	—	2.54	2.96 (13 mm)		0.4 P.F.A. Air. Water reducing	0.25 dia x 19	up to 3.3%

[a] Not stated whether total or free.
[b] Assumed that 1 bag cement = 90 lb.
[c] Added water.

estimate to be made of this fibre content for mixes containing aggregates of normal density (i.e. not lightweight aggregates)

$$W_f < \frac{600(1 - A_g)}{l/d} \tag{5.1}$$

where

W_f = weight of fibres, as a percentage of the concrete matrix, which can be compacted with normal site techniques

$$A_g = \frac{\text{weight of aggregate greater than 5 mm}}{\text{total weight of concrete}}$$

$$\frac{l}{d} = \frac{\text{length}}{\text{diameter}} \text{ of fibre}$$

Equation (5.1) is a simplification of that given in Reference 2.

5.5 MIXING METHODS

A variety of methods are available for introducing the steel fibres into the concrete mixer either to the dry constituents or to the wet mix. These techniques range from charging the aggregate conveyor with fibres[10,12] sieving fibres directly into the mixer drum,[7] sieving the fibres and blowing them into the drum[15] or alternating sieves for laboratory use.[2] Also, the development of fibres, glued with water soluble adhesive into units similar to staples,[16] enables the fibres to be dispersed into the mixer as a normal aggregate and they then separate in the mixing process.

The critical factor in whatever technique is used for single fibre addition is that the fibres should reach the mixer individually and be immediately removed from the point of entry by the mixing action. Mixer characteristics can affect the uniformity of fibre distribution but an excess of fibres can collect into balls in the mixer regardless of mixer type. Also, it is essential that no fibre balls are introduced into the mixer initially as these are unlikely to be broken up by the mixing action.

The fibres are normally delivered to the site in boxes or drums and equipment such as protective gloves, goggles, forks or rods may be useful to assist in transferring the fibres into the dispersing plant.

5.6 COMPACTION TECHNIQUES

Steel fibre concrete may be compacted by poker vibrators, by shutter or table vibration, or by surface vibrating beams as in floor slabs or slip form pavers. However, the type and direction of vibration can have a critical effect on the orientation of the wires relative to the future loading direction[2,5,17,18] and hence on the properties in the hardened state. Therefore, prototype products and laboratory or trial applications should be carried out using the same vibration techniques to be used in the full scale operation otherwise misleading strength performance could result.

Preferential orientation of the fibres under vibration may also be assisted by magnetic fields and increases in modulus or rupture of more than 50 per cent at 1.5 per cent by volume of fibres have been described by Bergström[19] using this technique. However, magnetic orientation is necessarily likely to be limited to precast applications under factory conditions.

The effect of fibre compaction techniques on the hardened properties is dealt with more fully in Chapter 6.

5.7 SPRAYED CONCRETE

Sprayed concrete, or gunite, has been successfully used with steel fibres in a variety of applications since 1971.[20] The traditional gunite techniques have been utilized with machines which have been modified so that the fibres can be mixed with the wet or dry constituents before being sprayed with or without additional water being added at the gun. Alternatively, the fibres may be separately projected into the wet or dry concrete as it is sprayed onto the work or they may be blown into the dry materials just before they emerge from the nozzle. The latter technique, which has been developed by the Besab Co in Sweden, is shown in Figure 5.4.

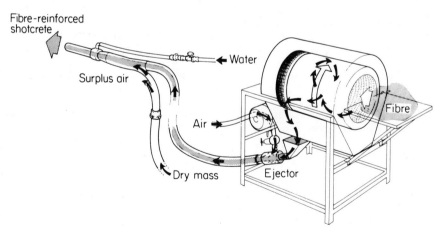

Figure 5.4 Fibre spraying equipment developed by Besab AB (Reproduced from Edgington,[20] *Conference on Fibre-reinforced Materials, 1977*, pp. 115–126, by permission of The Institution of Civil Engineers)

In Figure 5.4 the fibres are conveyed pneumatically from a rotary fibre feeder to the nozzle via a 75 mm diameter flexible hose. A particular advantage of the equipment is that it can be used with most standard dry shotcrete machines and it can handle fibres with length–diameter ratios up to 125.

REFERENCES

1. British Standards 1881: Part 2 (1970), *Methods of Testing Fresh Concrete*, British Standards Institution.
2. Edgington, J., Hannant, D. J., and Williams, R. I. T., 'Steel-fibre reinforced concrete,' *Building Research Establishment Current Paper CP 69/74*, July 1974.
3. Hannant, D. J., and Edgington, J., 'Fibre-reinforced concrete', *Conference Proceedings, Stevin Laboratory Delft*, 1974, pp. 63–70.
4. Mangat, P. S., and Swamy, R. N., 'Compactibility of steel-fibre reinforced concrete,' *Concrete*, 1974, 34–35, May.
5. Swamy, R. N., and Stavrides, H., 'Some properties of high workability steel-fibre concrete,' *Fibre-reinforced Cement and Concrete*, RILEM Symposium, 1975, Construction Press Ltd., pp. 197–208.
6. National Standard Co. Ltd., *Trade Literature*, Kidderminster.
7. Johnson and Nephew (Ambergate) Ltd., *Trade Literature on Wirand Concretes*, 1975, Derbyshire.
8. McDonald, A. R. 'Wirand concrete pavement trials. Fibrous concrete. Construction material for the seventies', *CERL Conference Proceedings M-28*, Champaign, Illinois 61820, December 1972, pp. 209–234.
9. Swamy, R. N., and Kent, B., 'Some practical structural applications of steel-fibre reinforced concrete,' *Fibre-reinforced Concrete*, American Concrete Institute Publication SP44, pp. 319–336.
10. Gregory, J., Galloway, J. W., and Raithby, K. D., 'Full-scale trials of a wire-fibre-reinforced concrete overlay on a motorway,' *Fibre-reinforced Cement and Concrete*, RILEM Symposium, 1975, Construction Press Ltd., pp. 383–394.
11. ACI Committee 544, 'State of the art report on fibre-reinforced concrete.' *American Concrete Institute Journal*, **70–65**, 729–744 (1973), November.
12. Gray, B. H., and Rice, J. L., 'Pavement performance investigation. Fibrous concrete: Construction material for the seventies,' *CERL Conference Proceedings M.28*, December 1972, Champaign, Illinois 61820, pp. 147–157.
13. Parker, F., 'Construction of fibrous concrete overlay, Tampa International Airport,' *CERL Conference Proceedings M.28*, December 1972, Champaign, Illinois 61820, pp. 177–197.
14. Johnston, C. D., 'Steel-fibre-reinforced concrete pavement. Construction and interim performance report,' *Proceedings of a Symposium on Paving, American Concrete Institute Convention*, San Francisco, April 1974 (Published as an ACI SP series, 1975)
15. McCurrich, L. H., and Adams, M. A. J., 'Fibres in cement and concrete', Current Practice Sheets, No. 5, *Concrete*, pp. 51–53 (1973), April.
16. Bekaerts, N. V. Ltd., *Trade Literature on Dramix®*, 8550 Swevegem, Belgium, 1975.
17. Edgington, J., and Hannant, D. J., 'Steel-fibre-reinforced concrete. The effect on fibre orientation of compaction by vibration,' *RILEM Materieux et Construction*, **5 (25)**, 41–44. (1972).
18. Hannant, D. J., and Spring, N., 'Steel-fibre-reinforced mortar: A technique for producing composites with uniaxial fibre alignment,' *Magazine of Concrete Research*, **26 (86)**, 47–48. (1974) March.
19. Bergström, S. G., 'A Nordic research project on fibre-reinforced cement-based materials,' *Fibre-reinforced Cement and Concrete*, RILEM Sympsoium, 1975, Construction Press Ltd., Vol. 2, pp. 595–600.
20. Edgington, J., 'Economic fibrous concrete.' *Conference on Fibre-reinforced Materials*, Institution of Civil Engineers, London, 1977, pp. 115–126.

Chapter 6
Steel-Fibre Concrete: Properties in the Hardened State

6.1 TYPES OF STEEL FIBRES

6.1.1 Fibre shape

It has been shown in Chapters 3 and 4 that the bond strength between the fibre and the concrete is one of the major factors which determines the properties of the hardened concrete. Manufacturers of fibre wire have attempted to improve the mechanical bond in a variety of ways and these have led to the different configurations shown in Figure 6.1.

Wires with a circular cross-section are produced by normal wire-drawing techniques which are relatively expensive, but cheaper ways of fibre production have been developed. The production method utilizing slit sheet results in rectangular section fibres which may be produced cheaply when supplies of scrap metal sheet are readily available and another economic technique is the 'melt extract' process* in which fibres are produced directly from the molten steel by means of a spinning multi-edged extraction disc in contact with the surface of the molten metal. In this process, the fibres are automatically thrown free of the disc giving the fibre shape shown in Figure 6.1 and cheap chromium steel and stainless steel fibres can be produced from scrap materials in addition to carbon steel fibres.

The wire types shown in Figure 6.1 have all been claimed by their manufacturers to give some extra benefits in bond when compared with plain round wire but there can also be disadvantages associated with the mechanical deformations. For instance, the indentation process used to make Duoform wire can weaken the wire and make it more brittle, particularly with the smaller diameter wires[1,2]. Also, stiff fibres combined with a sinusoidal shape can cause local bursting of the matrix due to radial tensile stresses exerted on the concrete as the fibres pull out. The ragged shape of fibres produced from the melt process is relatively untried but could cause

---

*Patented by the Batelle Development Corporation and developed in the U.K. by Johnson and Nephew (Ambergate) Ltd.

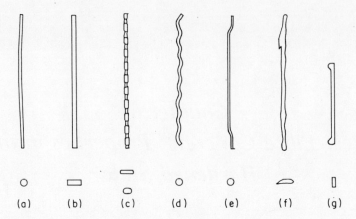

Figure 6.1. Shapes of steel fibres (a) Round, (b) Rectangular, (c) Indented (Duoform, National Standard Patent), (d) Crimped (G. K. N. and Johnson Nephew Ltd.), (e) Hooked ends (Dramix, Z. Bekaerto Ltd. Patent), (f) Melt extract process (Battelle Patent), (g) Enlarged ends (Australian Wire Industries Ltd. Patent)

problems if the bond is sufficiently good to cause fibre failure rather than pull out, because flexural strengthening relies on fibre slip occurring before failure (Chapter 4).

In addition to mechanically deforming the wire, various chemical and physical treatments have been tried in an attempt to improve the wire surface characteristics with respect to bonding with cement paste[2,3]. These treatments have included degreasing, surface roughening, and surface coatings, but even although the bond and pull out load of single wires can be considerably improved, the strength increases obtained by including treated wires in concrete are much less pronounced. The difference between single wire pull out tests and multiple wire pull outs has been investigated by Naaman and Shah[4] who have confirmed that the apparent bond strength is greatly reduced when many wires are involved.

6.1.2 Fibre strength

It is generally accepted that higher strength fibres tend to give higher strength composites in practice. However, this is not necessarily due to the higher failure stress of the wire because the majority of fibres pull out and the critical factor is the bond strength. The effect may therefore be related to the higher yield stress which can be utilized when the fibres are bent round sharp edges in the matrix as the cracks open out, a higher yield value absorbing more energy. Measured values of tensile strength on individual fibres have varied from 295 MN/m^2 to 2367 MN/m^2,[1] the lowest tensile strengths being observed on badly Duoformed wires.

6.2 CRACKING

The major benefits of steel fibre concrete have often been associated with the affect of fibres on the resistance of the concrete to cracking. A variety of cracking situations have been investigated and the claims for the material have included the following:

(a) Increases in the load at which visible cracks are formed under flexural loading.[5,6]
(b) Prevention of cracking in slabs subjected to shrinkage conditions[7] or limitation of cracking in highway pavements.[8]
(c) Increase in reinforcement stress for a given crack width when fibres are used in conjunction with mild steel or high tensile steel reinforcement in direct tension or flexure.[7,9,10]
(d) Resistance to crack propagation under fatigue loading.[11]
(e) Resistance to cracking under impact loading.[12]

The exact meaning of the term 'cracking' has been discussed in Section 3.1 and has been examined in detail in Reference 6. In practical applications the term is generally applied to cracks greater than about 0.05 mm i.e. just visible to the naked eye and there is no doubt that steel fibres can delay the appearance of cracking at this level. However, the fibres have been shown to have little effect on the crack initiation point at the microscopic level.[6] Claims (a), (d), and (e) above are discussed in following sections but (b) and (c) are covered briefly below.

6.2.1 Cracking in slabs

The length and width of cracks in highway pavements is an indicator of damage and hence of pavement life, and Johnston[8] has shown that slabs with 3.4 per cent by weight (one per cent by volume) of steel wires have considerably less cracking than equivalent plain concrete controls after a year of traffic loading. Also, Elvery[7] has shown that fibre reinforced slabs subjected to drying in a laboratory did not exhibit shrinkage cracking whereas plain concrete controls generally showed surface shrinkage cracks. The same specimens were then loaded, and the fibre slabs developed up to twice the number of cracks compared with the controls which allowed a doubling of the reinforcement stress for the same crack width.

6.2.2 Increase in reinforcement stress for a given crack width

The introduction of high yield reinforcing steels, combined with the crack width restriction of 0.3 mm given in C.P.110[13] has aroused interest in methods of increasing the working stress in the reinforcement while maintaining the crack width in the concrete within given limits. Steel fibres are beneficial in this respect in that they cause finer cracks at closer spacings because some load is carried by the wires across the cracked surfaces. This effect has been investigated in direct tension[7] and in flexure[7,9,10] and because the post-cracking flexural stiffness is increased with fibre concrete, deflections can also be somewhat reduced.

6.3 STRENGTH OF STEEL-FIBRE CONCRETE

6.3.1 Influence of fibre orientation

An important factor to be considered when the properties of steel-fibre concrete are quoted, is the direction of fibre orientation relative to the direction of applied stress. Although the fibres may be orientated randomly in three dimensions in the mixer, this is seldom the case after vibration and compaction have been completed and the hardened concrete can then exhibit anisotropic behaviour with strengths up to 50 per cent higher in one direction than another.[14,15]

Fibre alignment may be achieved accidentally or intentionally in a variety of ways. Table or surface vibration tends to cause the fibres to align in planes at right angles to the direction of vibration or gravity (Figure 6.2) and hence horizontal casting is preferable for beams or slabs.

Internal vibration causes a smaller amount of fibre alignment than table vibration and the effect of these two compaction techniques on the flexural strength of 100 mm x 100 mm x 500 mm beams has been measured,[14,15] the results of Swamy and Stavrides[15] being shown in Figure 6.3.

Component thickness is also a factor in fibre alignment and, as the emphasis is generally on thin sections for fibre concrete, this effect can be used to improve fibre efficiency. For instance, sections less than 100 mm thick containing 50 mm long fibres will inevitably result in the fibres tending towards 2-D rather than 3-D alignment.

Also, steel fibres may be aligned uniaxially either mechanically[16] or magnetically[17] with a considerable increase in strength of the composite provided that the stress is applied in a suitable direction.

Thus it can be seen that the strengths which are quoted in the literature for

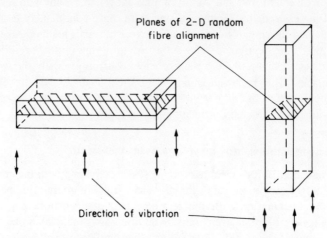

Figure 6.2. The effect of table vibration on fibre alignment (Reproduced from Edgington and Hannant,[14] *Materiaux et Structures*, 5(25), 41–44 (1972) by permission of RILEM

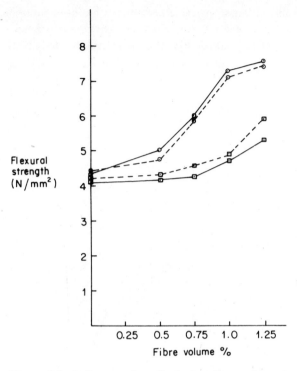

Figure 6.3. Influence of method of casting and type of vibration on flexural strength (Reproduced from Swamy and Stavrides,[15] *Fibre-reinforced Cement and Concrete,* 1, 206 (1975) by permission of the publishers, The Construction Press Ltd.); ——— External vibration; — — — Internal vibration; ⊙ Horizontal casting; ▫ Vertical casting

steel-fibre concrete are dependent on the test specimen size and shape and on the compaction procedure and this may explain some of the discrepancies noted between different authors.

From the practical point of view of manufacturing components, the anisotropic properties could be put to good use by arranging the compaction procedure so that the fibres are aligned in the most beneficial direction relative to the stress field. On the other hand, if the effects of vibration on fibre alignment are not fully appreciated, the strengths of steel-fibre-reinforced concrete products could be much lower than predictions based on laboratory tests using different compaction procedures.

6.3.2 Uniaxial tensile strength of composites containing random short steel fibres

The uniaxial tensile strength has been measured by several authors[1,6,18,19] for various fibre types and fibre volumes in matrices with particle sizes ranging from cement paste up to 20 mm aggregate concrete. It has been confirmed that at fibre volumes up to 3 per cent (10 per cent by weight) the maximum tensile strength increase is about 30 per cent and the maximum strengths rarely reach 5 MN/m² (see Figure 6.4). Increases in strain at failure are generally less than 50 per cent and the fibre spacing has little effect on these parameters. It is therefore not likely that steel-fibre concrete will be worth considering for applications where the tensile stresses are direct in type and are not able to be redistributed when cracking occurs.

6.3.3 Flexural strength

Surveys of flexural strength data have been made by Johnston,[20] Swamy[21] and the American Concrete Institute,[22] and extensive experimental work has been carried out by Lankard[23] and Edgington.[1] *All these authors are agreed that the major factors affecting the flexural strength are the volume fraction and the length/diameter ratio of the fibres, an increase in both these parameters leading to a higher flexural strength.* These conclusions are in agreement with the theory stated in Section 4.4.2 the other parameters being fibre orientation and bond strength There is a wide variability in absolute values for flexural strength in the published work possibly due to differences in matrix strength, bond strength, or fibre orientation, but the general conclusion is that flexural strength increases linearly with volume and length/diameter ratio of the fibres.

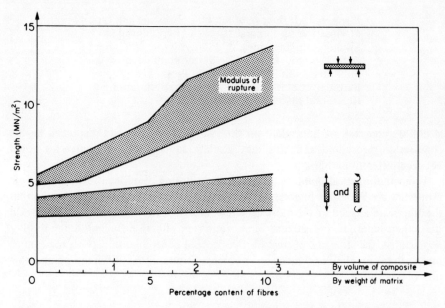

Figure 6.4. Flexural, direct tensile and torsional strengths of fibre-reinforced mortar and concrete (Reproduced from Edgington, Hannant, and Williams,[6] *Building Research Establishment Current Paper CP 69/74*, July 1974, by permission of The Controller, HMSO. Crown copyright reserved)

The effect of fibre orientation alone, on the theoretical modulus of rupture is shown in Figure 4.10 and the effect of bond strength alone for a 3-D fibre orientation is shown in Figure 4.11. When the effects of specimen size and test procedure are included with the other variables it is apparent that there can be no unique relationship between flexural strength and volume x length/diameter ratio of the fibres.

Moduli of rupture up to 10 to 15 MN/m^2 are possible using normal mixing and compaction techniques although a more commonly achieved value is about 7 MN/m^2. A summary of the results obtained by Edgington[6] is shown in Figure 6.4, the scatter band including mortars and concretes with a range of l/d ratios and wire strengths.

The difference between the moduli of rupture and direct tensile strengths for the same materials are also shown on Figure 6.4 and this difference is due to the apparent ductility and change in shape of the tensile stress block after cracking in flexure as explained in Section 4.2.

There is no doubt that the stress at first visual crack is also increased by the same parameters as the ultimate strength,[23] but the quoted values for first crack strength have a high variability because of the variety of techniques used to establish the cracking point, a major problem being the determination of the load at which the load-deflection curve deviates from linearity.

The shape of the load–deflection curve in flexure is examined in more detail in Section 6.4.

Poorly aligned fibres can give greatly reduced strengths (Figure 6.3) but if care is taken to align the wires uniaxially, flexural strengths up to 30 MN/m^2 can be achieved.[16]

The effect of beam depth and beam length can also alter the measured modulus of rupture[15] for the same material, deeper, longer beams tending to give lower strengths (Figure 6.5).

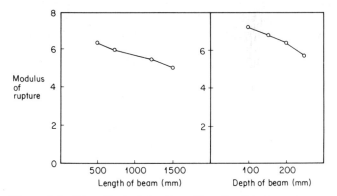

Figure 6.5. The effect of depth and length of test specimen on the flexural strength of fly ash fibre concrete (Reproduced from Swamy and Stavrides,[15] *Fibre-reinforced Cement and Concrete*, 1, 204 (1975) by permission of the publishers, The Construction Press Ltd.)

The large number of factors which can affect the modulus of rupture, therefore make it necessary to exercise considerable caution and to test trial products if the material is to be used in a structural or semi-structural situation.

6.3.4 Torsional strength

Plain concrete, when tested in torsion, fails by diagonal tension at about 45° to the axis. The calculated magnitude of the diagonal tensile strength depends whether elastic, semi-plastic or plastic theory is used for the calculation[24] and although elastic theory may be suitable for plain concrete the considerable post-cracking ductility of fibre concrete renders the plastic theory more appropriate. Edgington[1] carried out a detailed investigation into the torsional strength of steel fibre-reinforced mortar using square section prisms 100 mm x 100 mm x 500 mm long with fibre concentrations up to 4½ per cent by volume. The diagonal tensile strengths calculated on a plastic analysis were similar to the direct tensile strengths with increases over the control generally being less than 50 per cent, at 2 per cent by volume. However, a high proportion of the maximum torsional moment can be maintained at large rotations after cracking has occurred.

6.3.5 Fatigue strength

Reliable data on the fatigue of fibre concrete in flexure is rare and the following properties are from only one source[11] in which wires of length/diameter ratio between 70 and 90 were used at volume concentrations of 2 per cent and 3 per cent. The maximum stress applied in the fatigue cycle was expressed as a percentage of 'first crack stress' in static flexure which itself is ill-defined. However, the tests have shown that the fibre concrete will withstand 2×10^6 cycles at about 75 per cent of the first crack stress for fibre volumes of 3 per cent whereas at 2 per cent by volume of fibres the stress level drops to about 50 per cent for the same fatigue life.

Since the first crack strength is stated to be about 70 per cent of the modulus of rupture and the fibre volume in practical applications rarely reaches 2 per cent, considerable caution should be exercised before allowing for a greater fatigue strength in flexure for fibre concrete than for plain concrete.

Compression fatigue in fibre concrete has been studied by Ramey and McCabe[24(a)] who used 2 per cent by volume of 12.5 mm long by 0.15 mm diameter fibres and concluded that there was no apparent difference in the uniaxial compressive fatigue strength of fibre concrete and concrete without fibres.

6.3.6 Compressive strength

The compressive strength of cylinders and prisms has been measured for mortars and concretes[1] and it has been established that strength increases are generally less than 25 per cent for mortars at fibre volumes up to 4 per cent and about the same increase was measured for concrete at a fibre volume of 1.2 per cent. Similar results have been noted in a survey of published data by Johnston[20] and additional results published by Swamy[25] confirm that there is little point in including fibres in concrete to increase the compressive strength. However, there may be merit in including fibres to provide increased ductility in a compressive failure.

6.3.7 Strength in refractory applications

Although basic materials data is somewhat limited, refractory concretes reinforced with stainless steel fibres have proved effective in a variety of refractory applications at temperatures up to 1500 °C.[22,26,27,28] The properties required are resistance to large crack formation, spalling and abrasion, resistance to impulsive loads, and increased flexural strength.

Material properties have been measured by Nishioka et al.[28] on refractory castables after exposure to temperatures up to 1200 °C and improvements in flexural strength and energy absorption were observed at fibre concentrations between 5 per cent and 10 per cent by weight. Compressive strength was not affected. This particular material was used in a 250 mm thick layer on the door of a plate mill furnace and was successfully exposed to temperatures of 1270 °C for long periods.

6.3.8 Dynamic strength

Steel-fibre concrete is able to absorb a large amount of energy during fracture under dynamic loading. This is thought to be due to the substantial energy requirement to de-bond and pull out or yield and fracture the fibres as the cracks open at high loading rates. A measure of the energy absorption capability can be obtained from the area under the load-deflection curve in flexure as described in Section 6.4.

The dynamic properties can be utilized either to resist mechanical impacts or to reduce damage caused by shock waves such as explosive charges[29] or bubble collapse under cavitation conditions in spillways or sluices.[12]

An important factor in resistance to explosives is the fragment velocity and Williamson[29] showed for reinforced concrete slabs that the maximum fragment velocity could be reduced by 18 per cent with about 1¼ per cent wire by volume.

Hoff[12] has described steel-fibre concrete repairs to sluice ways, spillways, and spillway aprons in which cavitation had caused severe damage to plain concrete. The performance was stated to be excellent in comparison to plain concrete and was more durable than other repairs with polymer materials.

Mechanical impact testing has been carried out using a variety of techniques ranging from dropping full scale 39 tonne components,[12] Charpy tests on beams[7,30,31] blows on cubes with a ballistic pendulum,[32] drop ball tests on slabs,[15] or laboratory tests to simulate practical impact situations.[33]

The reported improvements in impact resistance with the inclusion of steel fibres vary widely and depend to a large extent on the energy and velocity of the

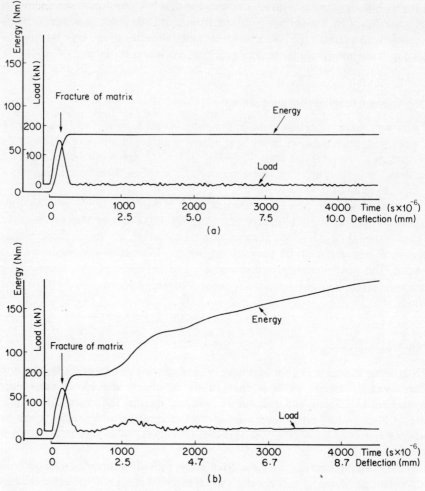

Figure 6.6. Typical load-time and energy-time curves for plain and steel fibre concrete subjected to a modified Charpy test (a) Plain concrete; MATRIX: OPC + 10 mm max. aggregate; $V_f = 0.0\%$; Age 2 months; Pendulum energy 250 N m; Velocity at impact 2.85 m s^{-1}; (b) Steel fibre concrete; MATRIX: OPC + 10 mm max. aggregate; FIBRE: Bekaert Ltd. DRAMIX, $l = 50$ mm, $d = 0.5$ mm; $V_f = 1.2\%$; Age 2 months; Pendulum energy 250 N m; Velocity at impact 2.85 ms^{-1}. (Reproduced from Hibbert and Hannant,[30] *Symposium, Sheffied*, April 1978, by Permission of A. Hibbert)

impacting mass, the size of specimen and rigidity of supports, the type of test, and even the definition of failure.

These types of machine effects are reflected in the literature where improvements of more than ten times the impact toughness of plain concrete have been reported but these are more likely to be functions of the test procedure than unique material descriptors.

One of the major benefits of steel fibres is the holding together of a cracked area after minor impacts, but the energy absorbed in separating a section into two pieces in a single impact can also be considerably increased.

An indication of the mechanism by which energy absorption may be achieved for 100 mm x 100 mm x 500 mm beams with a span of 400 mm subjected to a single blow fracture in a modified Charpy test[30] is shown in Figure 6.6.

In Figure 6.6 the energy has been obtained by integrating the load–time curves[30] and it can be seen for both plain and fibre concretes that the load reaches a similar maximum value in about 75×10^{-6} seconds. However, the load on the fibre concrete (Figure 6.6(b)) shows a second minor peak after about 1000×10^{-6} seconds which, presumably, results from the fibres carrying load as they extend, debond and pull out of the matrix. This second load hump is absent from the plain concrete but it makes a considerable contribution to the total energy to fracture as shown by the energy lines, that in Figure 6.6(a) for plain concrete being horizontal after 300×10^{-6} seconds whereas the energy absorption is continually increasing throughout a period of more than 4000×10^{-6} seconds for the fibre concrete shown in Figure 6.6(b).

The type and volume of steel fibre has also been shown to have a considerable effect on the energy absorbing characteristics of the composite[31] and hence, as with the other hardened properties, product testing is vital before the merit of fibre-reinforced concrete in impact situations can be fully assessed.

6.4 POST-CRACKING DUCTILITY

In practical applications, a factor which may be at least as important as an increase in the flexural strength, is the post-cracking ductility imparted by the fibres and this can be quantified by calculating the area under the load-deflection curve in slow flexure. (Figure 6.7)

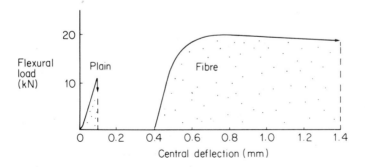

Figure 6.7. Load-deflection curves for plain concrete and steel fibre concrete. Beam size 100 mm x 100 mm Span 406 mm. 4 point load 1.5 per cent by Volume of fibres. $l/d = 100$ (Reproduced by permission of Edgington, *Ph.D. Thesis*, University of Surrey, 1973)

The area under the curve shown in Figure 6.7 is often described as a measure of the toughness or energy absorbing capability of the material and various values for toughness can be calculated for the same curve depending whether the complete load-deflection curve is used including the descending portion[19] or whether a cut-off point is chosen. From the point of view of serviceability of a structural unit a more meaningful value for toughness can be obtained from the area up to the maximum load or up to a specified deflection depending on the degree of cracking allowed in service. Johnston[20] has surveyed the data for the complete load-deflection curve and also for the area up to the maximum stress and Figure 6.8 has been plotted using the information in Reference 20.

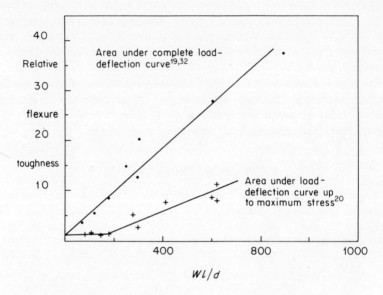

Figure 6.8 Influence of Wl/d on Flexure Toughness (Based on data from Johnston[20])

where $W = \dfrac{\text{weight of fibres} \times 100}{\text{weight of concrete}}$ l = fibre length
    d = fibre diameter

It can be seen that ability of the structural unit to absorb energy is substantial, even if the cut off point is taken at the maximum stress.

Another parameter which is of considerable practical importance and is also related to the load-deflection curve is the moment–rotation characteristic of the material. Thin sections of steel-fibre concrete can allow considerable rotation at a crack under a constant moment with a resulting redistribution of stress in a redundant system.

6.5 DIMENSIONAL CHANGES

6.5.1 Modulus of elasticity

The modulus of elasticity of concrete is largely controlled by the volume and modulus of the aggregate. Small additions of steel fibres would not be expected to greatly alter the modulus of the composite, as shown theoretically in Section 3.2, and this has been confirmed experimentally by Edgington et al.[6] Table 6.1 shows the effect of steel fibres on the modulus measured in direct tension for cement paste, mortar and 10 mm concrete.

Compressive stress–strain measurements on plain and fibre reinforced 10 mm concrete showed similar small increases in modulus, the values in tension and compression being essentially equal.

Table 6.1. Effect of fibres on tensile modulus of elasticity (Reproduced from Edgington, Hannant, and Williams, *Building Research Establishment Current Paper CP 69/74*, July 1974 by permission of The Controller, HMSO. Crown Copyright reserved)

Matrix	Volume of fibre reinforcement (%)	Tensile modulus of elasticity (average of 3 specimens) GN/m^2
Cement paste	0	26.4
	2.70	28.4
Mortar	0	33.9
	2.34	34.8
10 mm concrete	0	39.7
	1.47	40.9

6.5.2 Stress–strain curve

It has been shown in Section 3.4.8 that the volume of short steel fibres included in practice is generally less than the critical volume for fibre strengthening. Thus, the stress–strain curves in direct tension can be adequately represented by Figure 3.6(d), with the post-cracking strength generally being less than that of the matrix alone. The length of the falling load portion of the curve will depend on the fibre volume, bond characteristics, and crack widths.

However, it is possible to achieve the classic type of stress–strain curve shown in Figure 3.4 for a brittle matrix fibre composite by the use of continuous aligned steel wires in cement paste. Aveston et al.[35] have compared measured curves for fibre volumes of 2.3 per cent and 8.8 per cent with theoretical curves obtained using the theory by Aveston, Cooper, and Kelly.[36] The results are reproduced in Figure 6.9 and it can be seen that there is good agreement between theory and experiment.

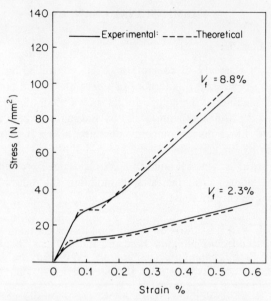

Figure 6.9 Tensile stress–strain curves for continuous steel wire reinforced cement (Reproduced from Aveston, Mercer, and Sillwood,[35] *Composites–Standards, Testing, and Design*, April 1974, pp. 93–103, by permission of the National Physical Laboratory. Crown copyright reserved)

6.5.3 Creep

From the results of compressive creep tests carried out over a loading period of 12 months[1] it has been found that the addition of steel fibres in concrete does not significantly reduce the creep strains of the composite. This behaviour is consistent with the low volume concentration of fibres when compared with an aggregate volume of approximately 70 per cent.

6.5.4 Shrinkage

The shrinkage of concrete over a period of 3 months on specimens subjected to various curing environments was unaffected by the presence of steel fibres. This again is as expected for the same reasons given for creep. However, the work described in Section 6.2.1 indicates that the pattern of cracking may be altered even although the overall movements may be similar.

6.6 DURABILITY

The liquid phase and initial hardened state in concrete made with Portland cement consist essentially of calcium and other alkaline hydroxides which form a highly alkaline material of pH between 12 and 13.[37] In this environment an insoluble oxide film forms on the surface of the steel wires which are protected from further corrosion while the film is unbroken.

However, atmospheric carbon dioxide dissolves in the moisture in the concrete to form weak carbonic acid which reacts with the alkalis in the pore water to form carbonates. This reduces the pH to values as low as 8.5 and, because the protective alkalinity of the concrete is then lost, corrosion of the steel can occur if oxygen and water are present.

The durability of steel fibre concrete is therefore largely dependent on the alkalinity of the matrix in the vicinity of the wires and this, in turn depends on the permeability of the concrete or more particularly on the extent of cracking.[38]

6.6.1 Uncracked concrete

Uncracked concrete, whether made with mixes recommended in CP110 1970[13] for marine exposure or with a porous lightweight aggregate with a high water/cement ratio has been shown to give good protection to the fibres for periods of 5 years in a variety of environments including industrial pollution and marine exposure.[39] The exposure trials were on 152 mm diameter x 308 mm long cylinders and although rust staining occurred on the surface of the concrete it was insufficient to cause serious spalling. Also, the carbonation depth was less than 5 mm after 5 years in the dense concrete (Figure 6.10) and occasional rusting up to depths of 6 mm was observed for the lightweight concrete.

Tests on uncracked flexural specimens have also indicated good durability with an increase in strength being measured after 5 years exposure with a temperature range of −23 °C to +38 °C and about 80 freeze–thaw cycles per year.[40]

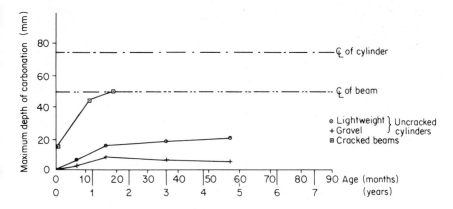

Figure 6.10 Relation between carbonation depth and duration of exposure for uncracked and cracked sections (Reproduced from Hannant and Edgington[39] *Fibre-reinforced Cement and Concrete*, 1, 165 (1975) by permission of the publishers, The Construction Press Ltd.)

Figure 6.11 Effect of carbonation and air borne marine spray on the corrosion of steel wires in pre-cracked beams. (a) Carbonation depth after 11 months is shown by lighter areas. Crack widths 6A–0.3 mm 6C–0.25 mm 6E–0.1 mm. (b) Rusting of 0.25 mm diameter wires at carbonated zone (11 months) (Reproduced from Hannant and Edgington,[39] *Fibre-reinforced Cement and Concrete*, 1, 168 (1975) by permission of the publishers, The Construction Press Ltd.)

6.6.2 Cracked concrete

Many of the proposed major applications of steel—fibre concrete are likely to involve flexural loading where the post-cracking ductility is probably a vital part of the load carrying capacity of the structural system. The durability in these cases may therefore be controlled by the rate of corrosion of the wires across the cracks.

Pre-cracked beams exposed to airborne marine spray have been used to study the carbonation rate within the cracks during periods of 19 months[39,41] The crack widths varied between 0.1 mm and 0.4 mm and the local carbonation rate was found to be greatly increased when compared with the uncracked cylinders as shown in Figure 6.10.

The depth of carbonation was dependent on crack width but had progressed to 15 mm from the surface after 11 months exposure in a tapering crack of maximum width 0.1 mm (Figure 6.11) and many of the fibres had started to corrode in the carbonated zone.

The surface area of the wire available for rusting was expected to affect the rate of wire damage and the loss in load carrying capacity of the beams and therefore two fibre sizes were used, 0.25 mm and 0.5 mm diameter. Although severe corrosion and extensive damage had occurred with many of the 0.25 mm diameter wires at 11 months and 19 months in the wider cracks (Figure 6.11), it was surprising to find that the failure load still generally exceeded the initial cracking load. This implied that a sufficient area of uncorroded fibre remained, which together with slightly increased bond, enabled the load capacity to be increased.

A theoretical study of corrosion damage[41] indicated that the 0.25 mm diameter wires could tolerate a reduction in cross-sectional area of about 75 per cent due to corrosion before the beam failure load would decrease and also the corrosion damage would take twice as long to become apparent from the beam failure load for the 0.5 mm diameter wires.

Thus, beam load capacity is not likely to be a good guide to corrosion damage in cracked fibre—concrete beams in the first few years of exposure.

Similar results, in which the load capacity of pre-cracked beams has increased up to 3 years exposure have been reported[40] but the five year results have shown a 20 per cent reduction in load capacity. This could indicate that wire corrosion has progressed sufficiently far to cause wire failure rather than wire pull-out.

In conclusion it may be assumed that fibre corrosion is likely to occur in cracked sections exposed to air and moisture but the rate of damage and loss of load capacity will depend greatly on the wire diameter and environmental conditions.

REFERENCES

1. Edgington, J., 'Steel-fibre-reinforced concrete', *Ph.D Thesis*, University of Surrey, 1973.
2. Tattersall, G. H., and Urbanowicz, C. R., 'Bond strength in steel-fibre-reinforced concrete', *Magazine of Concrete Research*, **V. 26 (87)**, 105—113 (1974) June.

3. Mayfield, B., and Zelly, B., 'Steel-fibre treatment to improve bonds', *Concrete*, **1973**, 35–37 (March).
4. Naaman, A. E., and Shah, S. P., 'Bond studies on orientated and aligned steel fibres', *Fibre-reinforced Cement and Concrete*, RILEM Symposium, 171–179, Construction Press Ltd., (1975).
5. Snyder, J. M., and Lankard, D. R., 'Factors affecting the flexural strength of steel fibrous concrete', *Journal American Concrete Institute*, **Title No. 69–9**, 96–100 (1972) February.
6. Edgington, J., Hannant, D. J., and Williams, R. I. T., 'Steel–fibre-reinforced concrete', *Building Research Establishment Current Paper CP 69/74*, July 1974.
7. Elvery, R. H., and Samarai, M. A., 'Reduction of shrinkage cracking in reinforced concrete due to the inclusion of steel fibres', *Fibre-reinforced Cement and Concrete*, RILEM Symposium, 1975, pp. 149–159.
8. Johnston, C. D., 'Steel-fibre-reinforced concrete pavement–second interim performance report', *Fibre-reinforced Cement and Concrete*, RILEM Symposium, 1975, pp. 409–418.
9. Swamy, R. N., and Al-Noori, K. A., 'Flexural behaviour of fibre concrete with conventional steel reinforcement', *Fibre-reinforced Cement and Concrete*, RILEM Symposium, 1975, pp. 187–197.
10. Hannant, D. J., 'Steel fibres and lightweight beams', *Concrete*, August 1972, pp. 39–40.
11. Batson, G., Ball, C., Bailey, L., Landers, E. and Hooks, J., 'Flexural fatigue strength of steel-fibre-reinforced concrete beams', *Journal American Concrete Institute*, November 1972, Title No. 63–64, pp. 673–677.
12. Hoff, G. C. 'The use of fibre reinforced concrete in hydraulic structures and marine environments', *Fibre-reinforced Cement and Concrete*, RILEM Symposium, 1975, pp. 395–407, Construction Press Ltd.
13. British Standard Code of Practice CP 110: Part 1 (1972), *The Structural Use of Concrete*.
14. Edgington, J., and Hannant, D. J., 'Steel-fibre-reinforced concrete: The effect on fibre orientation of compaction by vibration', *Materials and Structures*, RILEM, **5 (25)**, 41–44 January–February (1972).
15. Swamy, R. N., and Stavrides, H., 'Some properties of high workability steel fibre concrete', *Fibre-reinforced Cement and Concrete* RILEM Symposium, 1975, pp. 197–208, Construction Press Ltd.
16. Hannant, D. J., and Spring, N., 'Steel-fibre-reinforced mortar. A technique for producing composites with uniaxial fibre alignment', *Magazine of Concrete Research*, **26 (86)**, 47–48 (1974) March.
17. Bergström, S. G., 'A Nordic research project on fibre reinforced, cement based materials', *Fibre-reinforced Cement and Concrete*, RILEM Symposium, **2**, 505–600 (1975).
18. Johnston, C. D., and Coleman, R. A., 'Strength and deformation of steel fibre reinforced mortar in uniaxial tension', *Fibre Reinforced Concrete*, American Concrete Institute Publication, SP-44, pp. 177–193.
19. Shah, S. P., and Rangan, B. V., 'Fibre-reinforced concrete properties', *Journal American Concrete Institute*, Title No. 68–14, 126–135 (1971) February.
20. Johnston, C. D., 'Steel-fibre-reinforced mortar and concrete: A review of mechanical properties', *Fibre-reinforced Concrete*, American Concrete Institute Publication S.P. 44, pp. 127–142.
21. Swamy, R. N., Mangat, P. S., and Rao, C. V. S. K., 'The mechanics of fibre reinforcement of cement matrices', *Fibre-reinforced Concrete*, American Concrete Institute Publication S.P.44, pp. 1–28.

22. American Concrete Institute Committee 544, 'State of the art report on fibre-reinforced concrete', *A.C.I. Journal*, Title No. 70—65, 729—744 (1973) November.
23. Lankard, D. R., 'Flexural strength predictions', *Construction Engineering Research Laboratory, Conference Proceedings M-28*, December 1972, Champaign, Illinois 61820, pp. 101—124.
24. Zia, P., 'Torsion theories for concrete members', *American Concrete Institute Publication SP-18*, March 1968, pp. 103—133.
24 (a). Ramey, M. R., and McCabe, P. J., 'Compression fatigue of fiber-reinforced concrete', *Journal of Engineering*, Mechanics Division, American Society of Civil Engineers, **100**, No. EM2, April 1974.
25. Swamy, R. N., 'The technology of steel-fibre-reinforced concrete for practical applications', *Proceedings Institution of Civil Engineers, Paper No. 7694*, 1974, pp. 143—159.
26. Beckett, R. E., 'Patent positions and licence agreements in Europe', *Conference on Properties and Applications of Fibre-reinforced Concrete*, Delft University, 1974, pp. 187—190.
27. Lankard, D. R., and Sheets, H. D., 'Use of steel wires in refractory castables', *Ceramic Bulletin*, **50** (5), 497—500 (1971) May.
28. Nishioka, K., Kakimi, N., and Yamakawa, S., 'Effective applications of steel fibre reinforced concrete', *Fibre-reinforced Cement and Concrete*, RILEM Symposium, 425—433 (1975).
29. Williamson, G. R., 'Repsonse of fibrous reinforced concrete to explosive loading', *Technical Report 2-48* US Army Corps of Engineers, Ohio River Division Laboratories, January 1966.
30. Hibbert, A. P., and Hannant, D. J., 'The design of an instrumented impact test machine for fibre concretes', *Testing and Test Methods of Fibre-cement Composites*, RILEM Symposium, April 1978, Sheffield, England.
31. Hibbert, A. P., 'Impact resistance of fibre concrete', *Thesis submitted for Consideration for the Award of Ph.D.*, University of Surrey, 1978.
32. Dixon, J., and Mayfield, B., 'Concrete reinforced with fibrous wire', *Concrete*, 73—76 (1971) March.
33. Bailey, J. H., Bentley, S., Mayfield, B., and Pell, P. S., 'Impact testing of fibre-reinforced concrete stair treads', *Magazine of Concrete Research*, **27 (29)**, 167—170 (1975) September.
34. Blood, G. W., 'Properties of fibre reinforced concrete', *M.Sc. Thesis*, University of Calgary, July 1970, pp. 61—63.
35. Aveston, J., Mercer, R. A., and Sillwood, J. M., 'Fibre-reinforced cements — Scientific foundations for specifications. Composites, Standards, Testing and Design', *National Physical Laboratory Conference Proceedings*, April 1974, pp. 93—103.
36. Aveston, J., Cooper, G. A., and Kelly, A., 'Single and multiple fracture. The properties of composites', *National Physical Laboratory Conference Proceedings*, IPC Science and Technology Press Ltd., 1971, pp. 15—24.
37. Roberts, N. P., 'The resistance of reinforcement to corrosion', *Concrete*, October 1970, pp. 383—387.
38. 12-CRC Committee Report, Corrosion of reinforcement in concrete. State of the art Report', **12**, RILEM; Materiaux et Constructions, **9 (51)**, 187—206 (1976) May—June.
39. Hannant, D. J., and Edgington, J., 'Durability of steel-fibre concrete', *Fibre-reinforced Cement and Concrete*, RILEM Symposium, Construction Press, 1975, pp. 159—169.
40. Johnson and Nephew (Ambergate) Ltd., *Trade Literature*, 1975.

41. Hannant, D. J., 'Additional data on fibre corrosion in cracked beams and theoretical treatment of the effect of fibre corrosion on beam load capacity', *Fibre-reinforced Cement and Concrete*, RILEM Symposium, 1975, Volume 2, 1976. Contribution to paper 4.4, pp. 533–538.

Chapter 7
Polypropylene Fibres in Concrete, Mortar, and Cement

7.1 GENERAL

Polypropylene fibres were suggested as an admixture to concrete in 1965 by Goldfein[1] for the construction of blast-resistant buildings for the U.S. Corps of Engineers. His work comprised the incorporation of various natural and man-made fibres in mortar and neat cement and the publication gave the incentive for the early trials on polypropylene film fibre in concrete by Shell International Chemical Co. Ltd. who gave the material the name Caricrete.[2] The principles behind this early work have been described by Zonsveld.[3] Polypropylene twine is cheap, abundantly available, and like all man-made fibres, of a consistent quality so that few commercial or production problems were likely to arise if it could be shown to have potential in the field of concrete products.

It is felt necessary for a proper understanding of the practical aspects of using these fibres in concrete and mortar to explain briefly the manufacture and properties of the various types of polypropylene fibre.

7.2 THE MANUFACTURE OF POLYPROPYLENE FIBRES

The development of polypropylene in a new strong form, the isotactic configuration, and commercial production in the 1960s, offered the textile industry a potentially low-priced polymer capable of being converted into useful textile fibre. Polypropylene fibres then became available in two forms, monofilaments (or spinneret) fibres and film fibres. Extrusion of synthetic polymers into fibres by spinneret has long been the conventional method for rayon and nylon and this technique is also used to produce polypropylene fibres which are normally circular in cross-section and are used in many textile and carpet end uses.

The newer process of film extrusion is more economical and particularly suited for the processing of isotactic polypropylene. The extruder is fitted with a die to produce a tubular or flat film which is then slit into tapes, and monoaxially stretched. The 'draw ratio' is a measure of the extension which is applied to the fibre during fabrication and draw ratios of about 8 are common for polypropylene film. A molecular orientation results from the stretching and is the cause of the high tensile strength.

According to Spencer-Smith,[4] the structure of drawn polypropylene film can be described in terms of a model in which the polymer is itself a composite material containing a non-crystalline amorphous matrix reinforced by micro-fibrils formed from the deformation of small crystals (spherulites) during the hot drawing operation. Similar structures also occur in natural fibres such as linens, cottons, and wool.[5] This model of polypropylene film is identical to that of a fibre reinforced composite and hence the basic theory described in Chapter 3 can be used directly to explain the physical properties of the film.[4]

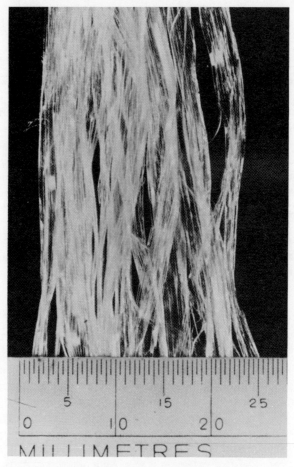

Figure 7.1. Fibrillated polypropylene film split by rubbing (Reproduced from Hannant, Zonsveld, and Hughes,[14] *Composites*, April 1978, by permission of IPC Science and Technology Press Ltd.)

Research into methods of drawing polyolefins with increased orientation and crystallization[6,7,8] may well lead to the commercial manufacture of polypropylene films with moduli of elasticity greater than the present market products.

Having achieved the production of films with adequate properties, their use in concrete is made possible by fibrillation which is the generation of longitudinal splits and can be controlled by the use of carefully designed pin systems on rollers over which the stretched films are led. A fibrillated yarn as shown in Figure 7.1 has been split by rubbing, resulting in random lengths over which the film is torn. Figure 7.2 shows the regular pattern of a pinned yarn. Fibrillated films twisted into

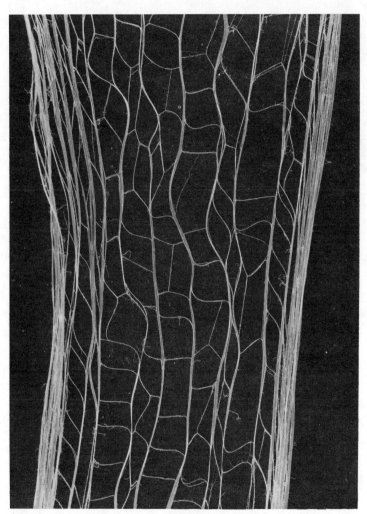

Figure 7.2. Fibrillated polypropylene film split by pin rolling (Reproduced from Hannant, Zonsveld, and Hughes,[14] *Composites*, April 1978, by permission of IPC Science and Technology Press Ltd.)

the form of fibres have a softer handle than spinneret fibres and were mainly developed for use in rope and twine, but proved to be useful in concrete as well.

The types of fibre are characterized by figures expressing the length in metres per kilogram, or by the old textile designation of the denier, the weight in grams of 9,000 m of yarn, For instance, a twine of 6,000 denier runs 1,500 m/kg.

The fibres are supplied in spool form for cutting on site, or are chopped by the manufacturer, usually in staple lengths between 25 and 75 mm. Purchasing the fibres on spools and cutting to the required length in the precast works can lead to considerable savings. Spools are also easier to handle in transport and require less storage space.

The flat fibrillated films may also be opened to form continuous networks which may themselves be impregnated with cement mortar to form composite sheets with high bending and impact strengths.

7.3 PROPERTIES OF POLYPROPYLENE FIBRES

The raw material polypropylene, derived from the monomeric C_3H_6, is a pure hydrocarbon like, for example, paraffin wax. According to Zonsveld[9] its mode of polymerization, its high molecular weight, and the way it is processed into fibres combine to give polypropylene fibres the following useful properties:

(a) A sterically regular atomic arrangement in the polymer molecule and high crystallinity. The regular structure gave it the name isotactic polypropylene.
(b) A high melting point (165 °C) and the ability to be used at temperatures over 100 °C for short periods.
(c) Chemical inertness making the fibres resistant to most chemicals. Any chemical that will not attack the concrete constituents will have no effect on the fibre either. On contact with more aggresive chemicals the concrete will always be the first to deteriorate.
(d) The hydrophobic surface, not being wet by a cement paste, helps to prevent chopped fibres from balling up during mixing like jute fibres. Or, stated in a different way, the water demand is nil.
(e) The stretching process as described in 7.2 results in a parallel orientation of the polymer chain molecules in the fibre, and in a high tensile strength, which in textile terms is 5 g/denier, equivalent to σ_{fu} = 400 MN/m^2
(f) The orientation leaves the films weak in the lateral direction, which facilitates fibrillation. The cement matrix can therefore penetrate in the mesh structure between the individual fibrils and create a mechanical bond between fibre and matrix.

Shortcomings are:

(a) Combustibility. A fire will leave the concrete with an additional porosity equal to the volume percentage of fibres incorporated in the case of chopped fibres, usually in the order of 0.3 to 1.5 per cent by volume.
(b) The low modulus of elasticity means that the inclusion of fibres reduces the

cracking strength of the composite (Equation 3.4) and results in very large strain before multiple cracking is complete (Figure 3.6).
(c) In respect of monofilaments only; the poor bond between fibre and matrix results in a low pull-out strength.
(d) Attack by sunlight and oxygen. To protect polypropylene against ultraviolet radiation and oxidation the manufacturers usually incorporate stabilizers and/or pigments, which result in fibres which are quite acceptable for use in ropes in a marine atmosphere. In addition, the surrounding concrete in the products to be discussed protects the fibres so well that this shortcoming is removed altogether.

The modulus of elasticity of the fibres ranges between 1 GN/m^2 and 8 GN/m^2 depending on the strain rate and is much lower than that of an average concrete, say 30 GN/m^2. Most materials, and plastics more than others, show a certain rate sensitivity, i.e. an increased rate of straining results in an increased modulus and Samuels[10] has indicated that the calculated high frequency dynamic modulus at sonic velocities, of polypropylene film could reach or exceed 15 GN/m^2, and measured results have exceeded 10 GN/m^2. This rate dependence of modulus for polypropylene may be significant when the impact strength of polypropylene concrete is considered and may be even more important for mortars where the matrix modulus may be about 20 GN/m^2.

7.4. PROPERTIES OF POLYPROPYLENE FIBRE COMPOSITES IN THE FRESH STATE

7.4.1. Mixing fibres into concrete[9]

A variety of mixers has been used in practice, some requiring an adjustment to the existing equipment, some none at all. Additional equipment has been installed in some plants to chop and/or to facilitate proportioning the fibres. The type of short fibre chosen is mostly based on film, e.g. a twine of 1400 m/kg, chopped to 50 mm staple length. As the fibres cannot be wetted, the mixing need only achieve a homogeneous dispersion and therefore they are often added shortly before the end of mixing the normal ingredients. A long residence time in any mixer leads to undesirable shredding of the fibres and should be avoided.

Tumbler mixers disperse the fibres without complications. This also applies to ready mix lorries which either carry a pre-weighed bag of fibres, or receive the fibres on site from stock held there. On arrival on site the fibres are dropped in the drum which is kept rotating for two or three minutes before placing.

Pan mixers, slow or high-speed, have sometimes needed adjustment to cope with fibres which have, of course, different dimensions from the normal aggregates. The scraper blades may need to be set to a different angle if the fibres are caught and collected on edges or in poorly streamlined corners. Also, the discharge opening may require widening if the fibres have tended to bridge the gate and clog it. Many pan mixers, however, have been found in practice to accept mixes with chopped polypropylene twine without alterations. In a fast Eirich pan mixer in one plant the mixing time for normal concrete was about one minute, and it was found that 0.5 per cent by volume of fibres could be added at the beginning without fear of shredding in this short mixing time.

Dry mixing cement, sand, and fibres for shotcreting has also proved possible without special precautions. Water in this operation was added at the gun orifice, and the fan which blew the dry mix through the hose worked without undue stoppages.

If the continuous twine or filament arrives on spools at the precast factory it is cut to a staple fibre by specially developed equipment. The cutter is placed in line with other batching machinery, and can also combine its task with accurate proportioning. At least two precast plants in the U.K. have such equipment installed. West's Piling and Construction chop twine of 700 m/kg to 40 mm long staple with a guillotine knife, while the fibre drops on to a running conveyor belt, which takes the other components of the mix to the pan mixer. The production rate of the cutter is accurately known so that a time switch starting and stopping its operation adds the exact amount of fibre per batch for a content of 0.44 per cent by volume in the piling shells. The other precast producer is John Laing, who use an air-powered cutter to chop bundles of polypropylene monofilaments, with individual diameters of 150 μm, to staple of 19 mm long for use in their 'Faircrete' products.[11]

The mix design of polypropylene concrete will take account the denier or runnage and the staple length of the fibre that will best suit the aggregate, the workability required and the equipment to be used in making the product. For instance, a thin-walled product would not accommodate the fairly stiff fibres of 700 m/kg because some would lie across the wall and would tend to break out on demoulding. The more flexible twine of 1400 m/kg would therefore be chosen, and would be cut to a shorter staple length. A heavy precast pile on the other hand would accept coarse fibres which would give a higher workability for the same fibre content.

If it is accepted that workability is partly dependent on the number and length of fibres in a unit volume of concrete it can be appreciated that a concrete cube of 100 mm side containing 63 fibres of 700 m/kg twine of length 50 mm is likely to have a higher workability than the same matrix containing 126 fibres of the same length of 1400 m/kg twine. This amount of fibre represents about 0.5 per cent by volume.

7.4.2 Mixing fibre with cement or mortar

Chopped fibres have been pre-mixed with cement paste at a water–cement ratio of 0.5 and on occasions polyethylene oxide has been used to assist the mixing process. After compaction in the mould the mixture has been de-watered by pressing and vacuum.[12]

Composites containing short fibres have also been made using the spray–suction techniques as used for glass fibre cements and fibre volumes up to 6 per cent (2.8 per cent by weight) have been achieved using this technique.[12]

Much higher fibre volumes, up to 12 per cent, are possible by the impregnation with fine mortar of networks of opened fibrillated film and combining a number of layers to produce thin sheets.

7.4.3 Workability of concrete containing polypropylene fibres

The workability of the mix is measured in standardized tests like the slump test, the V–B consistometer test and the compacting factor, which are all very relevant to fresh ordinary concrete, but are less so to a mix containing fibres. For instance, the slump of a mix with a low fibre content can be zero, although the mix flows satisfactorily when kept moving, and responds well to vibration.

Systematic work using the V–B test and compacting factor test with the fibres of different geometries as described in Section 5.3 for steel fibre concretes has not been carried out for polypropylene fibre concrete, but the experience in laboratory and plant is certainly in agreement insofar as a workability test should provide conditions of flow or vibration.

However, the effect of increasing fibre volume on the workability of a range of normal weight and light weight aggregate concretes has been studied by Ritchie et al.[15,16] who found that the compacting factor test does give a useful measure of the observed reduction in workability of the mix. Figure 7.3 is taken from

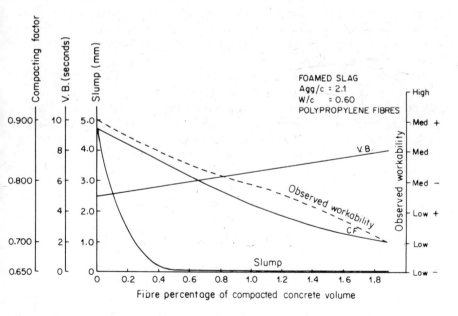

Figure 7.3. Comparison between the standard workability tests and the observed condition of the mix with increasing percentages of fibres (Reproduced from Ritchie and Al-Kayyali,[15] *Fibre-reinforced Cement and Concrete,* 1, 250 (1975) by permission of the publishers, The Construction Press Ltd.)

Reference 15 and shows the effect of increasing the volume of 35 mm long fibrillated film fibres of 1420 m/kg on the workability of a lightweight aggregate concrete.

Ritchie[16] also found that the progressive stiffening of polypropylene concrete mixes could be monitored by using the vane test. The shear strength figures so obtained showed not only the loss of workability with increasing fibre content but also the relative rate of increase of internal resistance with time.

In colloid–chemical systems like drilling muds, viscosity seldom follows the pure laws of viscous oils. The viscosity measurement gives different results depending on the rate of shear in the liquid. A vigorously stirred dispersion is seemingly less viscous than the slow-moving liquid, and the mud stiffens when stirring stops. This is termed thixotropy. The reverse, which is called dilatancy or rheopexy also occurs. These properties of suspensions and gels can also be encountered in fresh fibre/concrete composites, and can be used to advantage in developing new applications, or to streamline the production in the factory. If a mix needs an addition of plasticizer to enhance workability, the admixture may prove to have altered the rheological picture in a surprising direction and the mix composition or the operational procedure may have to be adjusted.[9]

John Laing Research and Development Ltd. have taken advantage of some of these novel properties of fibre concretes in the fresh state by stabilizing highly air entrained concretes (up to 40 per cent air by volume) with approximately 0.1 per cent by weight of chopped polypropylene monofilaments.[11] The main aim was to produce a range of concretes suitable for precast applications with improved thermal properties and with a decorative sculptured finish without the need for moulds and the new material was called 'Faircrete', which is a shortened version of fibre–air–concrete.

Variation of the amount of air entrainment or of the type of aggregate used allows a choice of ultimate densities from 700 up to about 2,000 kg/m^3. The polypropylene monofilaments have diameters ranging from 0.1 to 0.2 mm and are cut to lengths of 10 to 20 mm. Hobbs[11] likened the action of the fibres in the mix to that of a three-dimensional sieve, stopping the air passing up through the sieve and holding the aggregate so that it cannot pass down. The resulting properties of the mix, particularly when assisted with very light vibration, are easy flow out of hopper outlets, into restricted areas, and against mould faces. The thixotropic properties enable the concrete after placing to be formed into various shapes and patterns that would not be possible with ordinary concretes. The imprint does not slump back and remains exactly as formed on the hardened concrete.

For normal polypropylene fibre concrete, the way of handling fresh composites in the plant is dependent on the equipment available, and on the daily routine which was followed prior to the introduction of fibrous mixes. Alterations to the operational routine for normal concrete can eliminate all sorts of complications. Examples of changes implemented in practice have been: a wider discharge opening for the mixer drum, a conveyor belt in place of wheelbarrows and spades, a different external vibrator or mould shuttering, and other similar adaptations in the plant. Fortunately polypropylene concrete and mortar respond well to the conventional vibrating tables or pokers and presses, as the fibres, notwithstanding their low specific gravity, do not easily segregate from the mix.

Polypropylene fibres betray their presence usually on the finished surface, in most cases in an inconspicuous way, and not at all in cladding panels with exposed aggregates. Trowelling of panels and slabs poses problems for inexperienced factory hands, but a skilful operative can avoid the showing of fibres by dexterous use of his float.

7.5 PROPERTIES OF HARDENED COMPOSITES

Polypropylene fibres have been successfully used to increase the toughness of concrete subject to impact loading. However, the fibres have a low elastic modulus, a high Poisson's ratio and poor physicochemical bonding with cement paste and therefore have rarely been considered as promising fibres for reinforcing materials in direct tension or flexure or for the achievement of closely spaced multiple cracking in the relatively brittle matrix of cements and mortars.

Nevertheless, it has been shown theoretically[14] that there is no fundamental obstacle to increasing the load carrying capacity of a beam in flexure using polypropylene fibres and also that fine multiple cracking can be achieved at bond stresses well within the measured range for polypropylene and cement.

7.5.1 Bond strength

The desirability of achieving a good bond between fibre and matrix has been stressed repeatedly in the previous chapters and the means by which this can be achieved for short polypropylene fibres have been outlined by Zonsveld.[17] In terms of physicochemical adhesion there is no bond between polypropylene fibres and the cement gel. In fact polypropylene moulds for waffle floors are popular because of the ease of release after hardening. The use of the chopped and twisted fibrillated polypropylene fibres with their open structure has partially remedied the lack of interfacial adhesion by making use of wedge action at the slightly opened fibre ends and also by mechanical bonding through the fibrillations.

Tests have been conducted to measure the pull out loads of twisted fibrillated fibres[18,19] and these may range from 300–500 N for commonly used staples, but the accurate calculation of bond strength is complicated by a lack of knowledge of the surface area of fibre in contact with the paste. This area is likely to increase with time as the hydration products infiltrate the twisted films but approximate values of bond from Hughes' work[19] have been calculated to be 1.3 MN/m^2 for 12,000 denier fibres at 28 days. In practice, it is regularly observed in damaged products and in broken test specimens that the pull out strength of fibrillated polypropylene fibres is high and in specimens of an age of one year or more, fibre breakage instead of fibre pull out, is common.

Experiments to obtain a measure of the interfacial frictional bond for 150 μm diameter monofilaments[12] have indicated that values between 0.7 and 1.4 MN/m^2 can be obtained and that these are little affected by the environment in which the composite is placed although slight improvements in bond may occur with time. Some fibre breakage also occurred in the monofilament pull out tests.

The achievement of adequate bond is essential to ensure multiple cracking and for long fibres the crack spacing as a function of bond strength may be determined using the theory in Section 3.4.1 for round fibres or for rectangular sections as noted in Appendix I.

It has been shown[14] for a 30 μm film thickness at a fibre volume of 5 per cent that crack spacings of 5 mm can theoretically be achieved with bond strengths of only 0.23 MN/m^2. These crack spacings have been achieved experimentally by the use of opened networks of flat polypropylene film where the filaments are effectively continuous, and frictional bond is assisted by mechanical keying or pegging of the cement hydration products through the network and through slits in the film as shown in Figure 7.4

However, multiple cracking may not be possible with continuous smooth round aligned fibres at any volume loading because as Kelly and Zweben[20] have suggested, the fibre-matrix interface may debond in an unstable fashion due to the

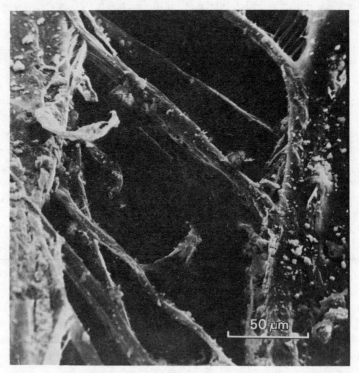

Figure 7.4. Film removed from cement matrix showing mechanical keying through slits in the film (Reproduced from Hannant, Zonsveld, and Hughes, *Composites*, April 1978, by permission of I.P.C. Science and Technology Press Ltd.

high Poisson contraction of the polypropylene at the crack. Thus, mechanical bonding or interlocking with the paste is probably essential if multiple cracking is to be achieved with aligned polypropylene fibres.

The slits in the film shown in Figure 7.4 may be at spacings of less than 10 μm if the film is in a state of imminent fibrillation.

7.5.2 Stress–strain curves in direction tension

A typical chopped-fibre composite containing about 0.44 per cent by volume of fibrillated twine, has an initial modulus similar to that of the matrix with a rapid drop in load after cracking accompanied by fibre pull out (see Figure 3.6d).

However, if the critical fibre volume for strengthening is just exceeded, for example by the use of opened networks of continuous film, then the type of stress–strain curve shown in Figure 7.5 can be achieved.[13] The slope of the curve in the post-cracking zone depends on the loading rate and fibre volume but even at the relatively low fibre volume of 2.3 per cent it is apparent that there is sufficient stress transfer from fibre to matrix to enable multiple cracking to occur as the composite strain increases. At higher fibre volumes the stress-strain curve in the

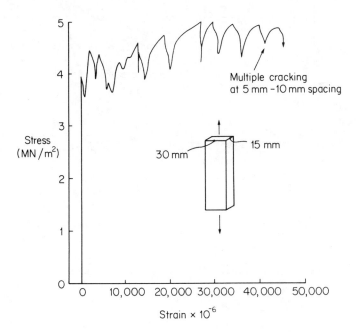

Figure 7.5. Tensile stress–strain curve for composite containing 2.3 per cent by volume of flat opened networks of polypropylene film (Reproduced from Hannant, Zonsveld, and Hughes, *Composites,* April 1978, by permission of I.P.C. Science and Technology Press Ltd.

post-cracking zone becomes much smoother and steeper and ultimate strengths in excess of 20 MN/m² are possible.

7.5.3 Static strength

For volume fractions of polypropylene less than one per cent it has generally been found[15,21,22,23] that increases in tensile, flexural, or compressive strengths of concrete are less than 25 per cent and often the strength of the composite is less than that of the matrix alone. This is basically due to the low modulus of elasticity of the fibres combined with less than the critical volume fraction.

On the other hand, if volumes of opened continuous film networks are included in cement paste or mortar, apparent moduli of rupture of more than 30 MN/m² can be achieved.[13,14] A rising type of load deflection curve in flexure is shown in Figure 7.6 for 6 per cent by volume of continuous film networks and this, combined with extensive multiple cracking as shown in Figure 7.7 enables large amounts of energy to be absorbed under impact loading.

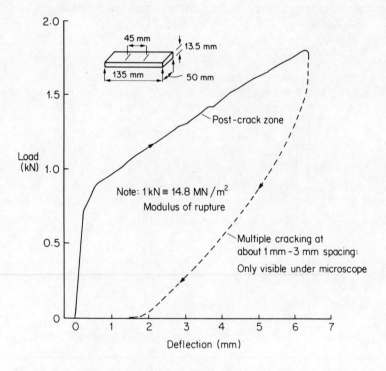

Figure 7.6. Load-deflection curve for a composite containing 6 per cent by volume of flat opened networks of polypropylene film (1 kN = 14.8 MN/m² modulus of rupture) (Reproduced from Hannant, Zonsveld, and Hughes,[14] *Composites*, April 1978, by permission of I.P.C. Science and Technology Press Ltd.

Figure 7.7. Closely spaced multiple cracking in flexure for a cement composite containing film fibres (Reproduced from Hannant, Zonsveld, and Hughes,[14] *Composites*, April 1978, by permission of IPC Science and Technology Press Ltd.

These properties may enable thin sheet products to be manufactured with adequate properties to act as replacements for asbestos cement sheets.

These properties may enable thin sheet products to be manufactured with adequate properties to act as replacements for asbestos cement sheets.

Moduli of rupture of up to 20 MN/m² have also been achieved with 2.8 per cent by weight (about 6 per cent by volume) of 170 denier monofilament using the spray process.[12]

7.5.4 Impact properties

The high impact and shatter resistance of cement composites reinforced with polypropylene fibres is partly due to the large amount of energy absorbed in debonding, stretching, and pulling out the fibres which occurs after the matrix has cracked.

However there is some evidence which indicates that even in the pre-visual cracking stage the impact toughness is improved. A factor which has not yet been quantified but may be beneficial, is the effect of small volumes of relatively soft fibres on the propagation of shock waves in the concrete

Improvements in impact strength from two to more than ten times that of the matrix have been measured in the laboratory by several workers[12,23,24] whose tests have included drop weights, the Izod pendulums, and explosive loading. However, quantitative impact strengths are of little value since they are highly dependent on the precise test procedure used. Comparative impact tests between

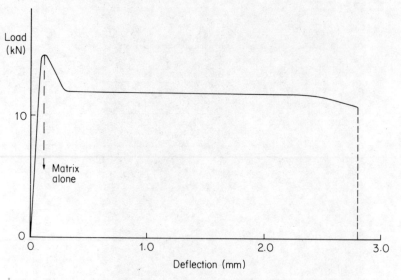

Figure 7.8. Load-deflection curve for 100 mm x 100 mm x 500 mm beam containing polypropylene chopped fibres (1.2 per cent by volume of fibrillated polypropylene, 700 m/kg, length 75 mm)

steel and polypropylene fibres using a modified Charpy machine have indicated that polypropylene can absorb as much energy as some steel fibres for the same fibre volume.[25,26]

Another measure of the ability to absorb flexural impact energy is the area under the load—deflection curve in slow flexure. Figure 7.8 shows such a curve for short chopped fibres at 1.2 per cent by volume and it can be seen that, compared with the plain matrix, the work to fracture is greatly increased even with a relatively low volume of 2D to 3D short random fibres.[9]

Commercial trials of polypropylene concrete products may give a better idea of likely field performance than laboratory work and this philosophy has been adopted by Wests Piling who use 40 mm lengths of fibrillated polypropylene twine at about 0.44 per cent by volume in concrete pile shells. The impact strength of the pile shells is assessed by dropping a hammer of 3 tonnes onto a shell taken from the production line. The end of a shell, without timber packing, is subjected to a series of hammer blows from increasing heights until failure is observed. Fairweather[27] has described details of these tests and the excellent resistance to impact of the shells, in comparison with steel mesh reinforced shells, has resulted in the manufacture of several million units since 1969. This is the largest single commercial application of fibre concrete as opposed to fibre cement up to 1978.

7.5.5 Fire resistance

At the Fire Research Station a panel of 0.9 m square and 50 mm thickness, containing 1.25 per cent by volume of polypropylene fibres, was subjected to a preliminary fire resistance test according to the 1967 British Standard 476 Part 1[28] and obtained a half hour rating. After fifty minutes in the furnace and at 930 °C on the exposed side, the maximum permissible temperature of 160 °C on the unexposed side was attained. Compared with concrete without fibres the presence of the fibre had made no difference to the vehaviour of the concrete under fire conditions.

The polypropylene fibres had melted during the test and, where the temperature had been high enough, the polymer had volatilized, leaving fine channels and an additional porosity in the panel, which was otherwise fully intact.

7.5.6 Durability

It is generally assumed, and has been confirmed by chemical tests, that there are no durability problems due to chemical degredation when polypropylene fibres are included in a cement matrix. However, mechanical test data on the tensile and impact strengths of the composite to substantiate this assumption are somewhat limited because the fibres have generally been included with concrete to give handling strength, or impact strength in the short term and long term strength has not been of great significance.

Some durability data have been obtained by Walton and Majumdar[12] for composites containing short lengths (20 mm—50 mm) of 170 denier monofilament at up to 2.8 per cent by weight (6 per cent by volume) and also for short lengths of fibrillated film fibre of 1000 denier and 12,000 denier at up to 2 per cent by weight (4.4 per cent by volume). Curing conditions have included water, air, natural weather, and accelerated curing under water at 60 °C for one year, which is considered to be a severe test for this type of material.

The general conclusions from this work[12] were that there was little change in modulus of rupture or impact strength with time under these storage conditions and that the accelerated testing results indicated that the high impact strength derived from polypropylene will remain stable over very long periods of time in normal use.

7.6 APPLICATION OF PHYSICAL DATA TO THE DEVELOPMENT OF COMMERCIAL PRODUCTS

Existing standard testing methods for concrete measure more the raw material properties than the behaviour of the finished commercial product. The previous sections have indicated that polypropylene fibre concrete as a material, fresh or cured, behaves differently from conventional concrete and therefore, the translation of raw material properties to the properties of the final article is more difficult than it is for plain concrete. This difficulty is even greater than for other types of fibre, because the support for measured properties by theoretical considerations is less well developed for low modulus fibre concretes than it is for high modulus fibre composites. It is therefore recommended that tests of finished products be given preference in judging the merits of polypropylene concrete for new applications.

Examples of product test specifications are the British Standard Specifications for concrete pipes, BS 556,[29] which includes hydraulic and crushing tests of pipes, fittings, manholes, and gullies, and BS 368,[30] for precast flags, which specifies a test for transverse strength by centre-line loading and one for water absorption.

Polypropylene products are likely to require greater care in compaction than other fibre concretes because a characteristic of fibrillated twine is that it acts as a damper or energy absorber during the compaction process. Density checks on the compacted concrete should therefore be made at intervals as it has been found in the past that poor results could often be explained by increased porosity due to poor compaction.

REFERENCES

1. Goldfein, S., 'Fibrous reinforcement for Portland cement,' *Modern Plastics*, **1965**, 156–159 (April).
2. Zonsveld, J. J., and Salmons, R. F., *British Patent Specification, No. 1, 130, 612* (1968).
3. Zonsveld, J. J., 'The marriage of plastics and concrete,' *Plastica*, **23**, 474–484 (1970) October.
4. Spencer–Smith, J. L., 'Microfibril–matrix theory of drawn polypropylene,' *Plastics and Rubber: Materials and Application*, **1976**, 74–80 (May).

5. Hearle, J. W. S. 'The structural mechanics of fibres,' *Journal of Polymer Science: Part C*, **20**, 215–251 (1967).
6. Capaccio, G., and Ward, I. M., 'Preparation of ultra-high modulus linear polyethylenes; effect of molecular weight and molecular weight distribution on drawing behaviour and mechanical properties,' *Polymer*, **15**, 233–238 (1974) April.
7. Capaccio, G., Crompton, P. A., and Ward, I. M., 'Ultra-high modulus of elasticity polyethylene by high temperature drawing,' *Polymer*, **17**, 644–645 (1976).
8. Zwijnenburg, A., and Pennings, A. J., 'Fibrillar polyethylene crystals,' *Journal Polymer Science, Polymer Letters Edition*, **14**, 339–346 (1976).
9. Zonsveld, J. J., 'Polypropylene fibre concrete,' *Fibre Cement and Fibre Concrete; Course Notes for Practising Civil Engineers*, University of Surrey, April 1976.
10. Samuels, R. J., 'Spherulite structure deformation morphology, and mechanical properties of isotactic polypropylene fibres, *Journal of Polymer Science: Part C*, **20**, 253–284 (1967).
11. Hobbs, C., 'Faircrete: an application of fibrous concrete. Prospects for fibre reinforced construction materials,' *Proceedings of International Building Exhibition Conference*, November 1971, Published by Building Research Establishment, pp. 59–67.
12. Walton, P. L., and Majumdar, A. J., 'Cement-based composites with mixtures of different types of fibres," *Current Paper CP 80/75*, Building Research Establishment, September 1975.
13. University of Surrey and Hannant, D. J., *Provisional Patent Specification*, No. 27371/76, July 1976.
14. Hannant, D. J., Zonsveld, J. J., and Hughes, D. C., 'Polypropylene fibres in cement based materials,' *Composites*, **9**, No. 2, April 1978, I.P.C. Science and Technology Press Ltd., pp. 83–88.
15. Ritchie, A. G. B., and Al-Kayyali, O. A., 'The effects of fibre reinforcements on lightweight aggregate concrete,' *Fibre-Reinforced Cement and Concrete*, RILEM Symposium, 1975, Construction Press Ltd., pp. 247–256.
16. Ritchie, A. G. B., and Rahman, T. A., 'Effect of fibre reinforcement on the rheological properties of concrete mixes,' *Proceedings International Symposium on Fibre-reinforced Concrete*, Ottawa, October 1973. *Properties of Fibre-reinforced Concrete*, ACI Publication, SP 44 1974, Part 1, pp. 29–44.
17. Zonsvield, J. J., 'Properties and testing of concrete containing fibres other than steel,' *Fibre-reinforced Cement and Concrete*. RILEM, Symposium, London, 1975, Construction Press Ltd., pp. 217–226.
18. Ritchie, A. G. B., and Mackintosh, D. M., 'Selection and rheological characteristics of polypropylene fibres,' *Concrete*, pp. 36–39 (1972) August.
19. Hughes, B. P., and Fattuhi, N. I., 'Fibre-reinforced concrete in direct tension,' *Fibre-reinforced Materials*, Paper 16, Conference at Institution of Civil Engineers, London 1977, 127–133.
20. Kelly, A., and Zweben, C., 'Poisson contraction in aligned fibre composites showing pull out,' *Journal of Materials Science*, **11**, Letters (1976), pp. 582–586.
21. Dardare, J., 'Contribution to the study of the mechanical behaviour of concrete reinforced with polypropylene fibres,' *Fibre-reinforced Cement and Concrete*, RILEM Symposium, London 1975, pp. 227–236.
22. Majumdar, A. J., 'Properties of fibre–cement composites,' *Fibre-reinforced Cement and Concrete*, RILEM Symposium, London 1975, pp. 279–314.
23. Nanda, V. K., and Hannant, D. J., 'Fibre-reinforced concrete,' *Concrete Building and Concrete Products*, Vol. XLIV. No. 10, 179–181 (1969) October.

24. Raouf, Z. A., Al-Hassani, S. T. S., and Simpson, J. W., 'Explosive testing of fibre-reinforced cement composites,' *Concrete,* 28–30 (1976) April.
25. Hibbert, A. P., and Hannant, D. J., 'The design of an instrumented impact test machine for fibre concretes,' *International Symposium on Testing and Test Methods of Fibre Cement Composites*, University of Sheffield, April 1978, Paper 8.
26. Hibbert, A. P., 'Impact resistance of fibre concrete,' *Thesis submitted for consideration for the Award of Ph.D.*, University of Surrey, 1978.
27. Fairweather, A. D., 'The use of polypropylene fibrillated fibres to increase impact resistance of concrete,' *Proceedings of an International Building Conference*, Prospects for fibre-reinforced construction materials, November 1971, pp. 41–44, Published by the Building Research Establishment.
28. British Standards 476: Part 1 (as in force in 1967). Altered now under the title British Standards 476: Part 4, *Non-combustibility Test for Materials.*
29. British Standards 556: (1972), *Concrete Cylindrical Pipes and Fittings.*
30. British Standards 368: (1971), *Precast Concrete Flags.*

Chapter 8
Glass Fibres in Cement and in Concrete

8.1 BACKGROUND

Much of the credit for the initial development of dispersed glass-fibres as reinforcement for cement must undoubtedly be given to the Russians, Biryukovich et al. who published a book[1] in 1964 on the properties and basic methods of fabrication of the material. Biryukovich had previously been involved in the construction of a factory roof in Kiev in 1963 and early work had also been carried out in China since 1958.

The Russian work was mainly concerned with binders of low alkali or high alumina cement and this stimulated the work of Majumdar and Nurse at the Building Research Establishment in the U.K. which was directed towards the development of a glass fibre which would resist attack by the highly alkaline ordinary Portland cements commonly used in Europe and America.

The identification of alkali-resistant glasses by the Building Research Establishment and their subsequent development and commercial production in the U.K. by Pilkington Brothers Ltd. lead to major research efforts by both the B.R.E. and Pilkington in the early 1970s directed towards understanding the physical properties of the composite material and this resulted in the publication of a large amount of data of value to the construction industry, much of which is contained in References 2 to 8.

The use of a material in structural situations must rely on adequate material properties being maintained for periods up to 50 years and therefore much research effort has been devoted to durability testing and to the prediction of long term properties. However, the confirmation of predicted values can only be achieved by tests on the materials or components after continuous exposure to the appropriate weathering conditions and hence the material can not be proved suitable for continuously loaded structural applications until many years have elapsed.

8.2 MANUFACTURE OF GLASS FIBRES

Glass wool with fibre lengths up to 150 mm can be produced by blowing compressed air or steam at a stream of molten glass and similar but longer fibres can

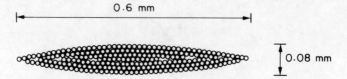

Figure 8.1. Idealized view of a glass-fibre strand containing 204 monofilaments each of a diameter of 10 μm (Reproduced from Krenchel,[9] *Fibre-reinforced Cement and Concrete*, **1**, 77 (1975) by permission of the publishers, The Construction Press Ltd.)

be manufactured by centrifuging molten glass. However, these fibres, although cheap are not very suitable for mixing with cement and therefore mechanically drawn fibres are used for this purpose.

Mechanical production of continuous fibres consists of drawing filaments from the bottom of a heated platinum bushing or tank containing several hundred holes. The glass fibres are collected in strands of about 200 filaments on a rotating drum and their final diameter depends on the speed of rotation of the drum, the viscosity of the melt and the size of the holes in the bushing.

The collection of 200 or so filaments is known as a strand but before reaching

Table 8.1. Chemical composition of some glasses (weight per cent) (Reproduced from Larner, Speakman, and Majumdar,[10] *Journal of Non-crystalline Solids*, **20**, 43–74 (1976) by permission of North-Holland Publishing Company)

Composition	A-glass	E-glass	G20 glass (Zirconia glass)
SiO_2	72.66	54.50	70.27
Fe_2O_3	0.13	0.27	0.05
TiO_2	0.05	0.51	0.07
Al_2O_3	1.10	14.22	0.24
ZrO_2	–	–	16.05
CaO	8.43	17.32	0.04
MgO	3.85	4.71	≯0.04
Mn_2O_3	0.01	0.01	–
Na_2O	12.81	0.32	11.84
K_2O	0.57	0.16	0.04
Li_2O	–	–	1.04
SO_3	0.23	–	≯0.02
B_2O_3	–	7.94	–
As_2O_3	–	–	0.04
PbO	–	–	0.03
Total	99.84	99.96	99.77

the drum, the strand is coated with a size which holds the filaments together in a lens shaped form as shown in Figure 8.1

The cross-sectional area (A_f) and perimeter (P_f) of the strand have been estimated to be 0.016 mm^2 and 2.64 mm respectively by Krenchel[9] and 0.027 mm^2 and 1.42 mm by Ali et al.[6] Oakley and Proctor[9a] have shown for strands of 204, 12.5 μm fibres about 0.65 mm wide by 0.11 mm thick that $A_f = 0.074$ mm^2 and $P_f = 2.83$ mm. This means that the glass occupies 34 per cent of the strand volume.

Several strands may be lightly bonded to form a roving which may be wound as a 'cheese'. The roving can be later unwound from the inside of the cheese to be chopped to suitable lengths for use directly in cement or for the production of chopped strand mat.

Mechanically drawn glass fibres are commonly available in three groups, soda-lime-silica glass or A-glass; borosilicate glass or E-glass; and Zirconia glass which is more resistant to attack by alkalis than A or E glass. Typical chemical compositions for these glasses are given in Table 8.1.

8.3 PRODUCTION TECHNIQUES FOR GLASS FIBRE-CEMENT BASED COMPOSITES

8.3.1 General

Most of the commercially attractive methods of producing flat sheet materials from glass fibres and cement paste are developments of those used by Biryukovich,[1] who in turn followed the technology used in the glass reinforced plastics industry.

The use of chopped fibres in mortar or concrete is also feasible but the mixing techniques for these materials are similar in many respects to those described for steel fibres and polypropylene fibres (Sections 5.5 and 7.4.2 respectively).

For sheet materials the mixing, fabrication, and compaction procedures are very important because the final orientation of the fibres relative to the future direction of applied stress is controlled by the procedure adopted. Thus a decision should only be taken regarding a fabrication technique after consideration has been given to the expected stresses in the product during handling and service conditions.

8.3.2 Premixing[4]

Premixing is a process in which all the constituents, including short strands of glass fibre are intimately mixed and then further processed to produce a product by casting in open moulds, pumping into closed moulds, extrusion, or pressing. The orientation of fibres tends to be three dimensional random in the mixer but may be altered to a limited extent by the production process.

Glass fibres tend to tangle and matt together if care is not taken in the mixing process and generally pan mixers have been found to give a better result than drum mixers. It has been suggested that it is preferable to disperse the fibres in water containing a thickening additive such as polyethylene oxide or methyl cellulose (0.1 per cent to 1 per cent of total mix water) before adding the solids and mixing in a Cumflow type of mixer but stiff fibre strands have also been produced for the premix process which make the use of thickening admixtures unnecessary.

Typically, fibre contents would be between 2 per cent and 5 per cent by weight of the other dry materials using chopped strands about 25 mm long. The highest fibre contents cause some compaction difficulties but the simplest methods of hand tamping in 25 mm layers followed by mould vibration can be effective. The four main methods of production of products from premixed material are described below.

8.3.3 Gravity moulding

Although the casting process is similar to some normal precast concrete production techniques, the glass reinforced premix allows the use of much thinner sections in both open or double sided moulds. Poker vibration is not recommended and generally external mould vibration is sufficient to produce flow, although the mix is less mobile than most concretes. Decorative aggregate finishes can be applied as for normal precast concrete.

8.3.4 Pressing

Flat sheets between 10 mm and 20 mm thick have been produced by pressing premix at pressures between 0.15 MN/m^2 and 10MN/m^2. Vacuum assisted dewatering using paper-felt filters may be from either one or both mould faces but a water thickening admixture is required to prevent water expulsion before the fibrous mix has had time to uniformly fill the mould.

A typical mix may contain ordinary Portland cement, a filler such as pulverized fuel ash and zone 3 sand at a water/cement ratio, before pressing of about 0.8. Proportion by weight of the solids may be O.P.C: P.F.A: sand of 1:1:1 with small quantities of polyethylene oxide or methyl cellulose admixture. Mixes of this type can accept between 1.7 per cent and 2.5 per cent (by weight of dry solids) of glass fibres with lengths between 11 mm and 22 mm.

Simple components with a good surface finish have been produced using this technique and immediate de-moulding is possible.

8.3.5 Injection moulding

Premix containing a thickening admixture and with up to 5 per cent of glass fibre by weight of dry solids has been successfully pumped into closed moulds under pressure. Fibre damage may be increased by the pumping process but window frames, fence posts, and hollow columns have been cast using this process. However, care is required if the presence of small blow holes on the surface of the product are to be avoided.

8.3.6 Extrusion

Extruded sections with complex shapes have been produced commercially but careful attention is required to mix design to prevent bleeding through the mix or blocking of the die.

Vibration can be used to assist the flow of the material and it is possible to arrange the vibration and extrusion process to give a beneficial fibre alignment in the finished product.

8.3.7 Spray Techniques

Spray Suction

This technique has been developed from the glass-reinforced plastics industry and consists of leading a continuous roving up to a compressed air operated gun which chops the roving into lengths of between 10 mm and 50 mm and blows the cut lengths at high speed simultaneously with a cement paste spray onto a forming surface containing a filter sheet and the excess water can then be removed by vacuum.

The fibre-cement sheet can be built up to the required thickness and demoulded immediately, an additional advantage being that it has sufficient wet strength to be

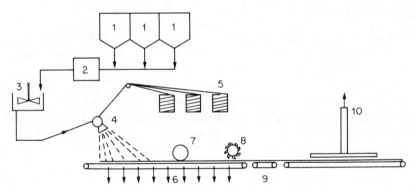

Figure 8.2. Main stages in the automated production of glass fibre reinforced flat or corrugated sheets using the spray-suction process

1. Silos
2. Weigh batcher
3. Mixer
4. Traversing head spraying glass fibre and cement slurry
5. Bobbins with glass-fibre roving
6. Vacuum dewatering
7. Roller finisher
8. Side and cross cut saws
9. Conveyors
10. Vacuum lift

bent round radiused corners to give a variety of product shapes such as corrugated or folded sheets, pipes, ducts, or tubes. Coloured fine aggregate can be trowelled into the surface to produce a decorative finish.

The process can be readily automated for standard sheets as shown in diagrammatic form in Figure 8.2.

Direct Spray

In principle, this process is similar to the spray suction technique except that the addition of suitable admixtures to the matrix reduces the water requirement of the cement slurry for satisfactory spraying and therefore the filter sheet and suction operation can be dispensed with. Roller compaction is often used to ensure compliance with the mould and to assist in the removal of entrapped air. Water/cement ratios as low as 0.30 to 0.35 may be achieved using this technique for neat Portland cement.

The density may be slightly lower for direct spray materials, lying in the range 1750 to 2000 kg/m^3 as compared with 2000 to 2100 kg/m^3 for spray–suction materials.[8]

The direct spray technique is probably the most used for building components of complex shapes such as permanent formwork or cladding panels.

8.3.8 Winding process

Figure 8.3 shows the main elements in the winding process for the production of pipes although similar procedures can be adopted for sheets or open sections by cutting and re-forming the freshly made composite.

The continuous glass fibre rovings are impregnated with cement slurry by passing them through a cement bath and they are then wound onto a suitable mandrel at a predetermined angle and pitch. Additional slurry and chopped fibre can be sprayed onto the mandrel during the winding process and roller pressure combined with suction can be used to remove excess cement paste and water. Fibre volumes in excess of 15 per cent have been achieved and hence very high strengths are possible. The process, as for the spray suction process, can be readily automated for production runs.

8.3.9 Lay-up process

The continuous impregnated rovings used in the winding process can also be laid in moulds in the form of window frames or similar products which can then be vibrated or pressed to improve the penetration by cement paste.

Hand lay ups using random glass fibre mat or continuous fibre fabric can also be produced on surface moulds on which the impregnation of cement slurry is assisted by surface rollers, pressure, and suction. The manufacture of complex shapes of the types produced in glass-reinforced plastics is possible using this technique.

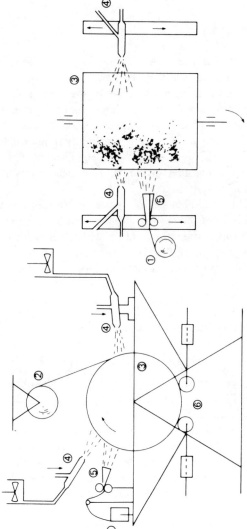

Figure 8.3. Fabrication of glass-fibre reinforced pipe using the winding process

1. Bobbin
2. Roving
3. Winding cylinder
4. Cement paste sprayer
5. Chopped fibre sprayer
6. Compression rollers

(Reproduced from Majumdar,[2] *Building Research Establishment Current Paper CP 57/74*, May 1974, by permission of The Controller, HMSO. Crown copyright reserved)

8.4 GLASS FIBRES IN HIGH ALUMINA CEMENT

Most applications of glass-reinforced cement are likely to utilize ordinary Portland cement but much of the early work with E-glass fibres by Biryukovich in Russia and by 'Elkalite Ltd' in the U.K. was in conjunction with high alumina cement and therefore the properties of this composite are reviewed briefly herein.

8.4.1 High alumina cement

The original reason for using high alumia cement with E-glass fibres was that it is a cement with a lower alkalinity than ordinary Portland cement and it was therefore thought to be less likely to cause chemical attack of the glass the O.P.C. The pH of the aqueous phase in contact with set high alumina cement is 11.8–12.05 with a total alkali content of 0.15 to 0.2 per cent and no calcium hydroxide is formed on setting, the strength being provided by hydrated calcium aluminates.[3] In contrast, the pH of the aqueous phase in equilibrium with set Portland cement is 12.5–13 with a considerable amount of well crystallized calcium hydroxide being formed during hydration.

However, storage of the set high alumina cement paste under warm moist conditions alters the chemical structure of the reaction products resulting in a paste with increased porosity and decreased strength. Contact with alkaline or acid solutions can further reduce the strength of the converted paste and, since the mid-1970s, high alumina cement has not been recommended for use in structural situations in the U.K.

8.4.2 Properties of glass fibre-high alumina cement composites

8.4.2.1 Direct tension

The basic properties in tension were determined by Biryukovich[1] who obtained the stress–strain curves shown in Figure 8.4 for continuous E-glass fibres aligned parallel to the direction of load application.

Figure 8.4 shows that the failure stress increases with glass content and that very high stresses and strains can be achieved, the tensile strength at 9 per cent by weight of glass being 60.8 MN/m^2. Biryukovich also found that the tensile strength was little affected by changes in the cement grade or w/c ratio and later theoretical work (Section 3.4) showed that the shapes of the stress–strain curves are in agreement with theoretical predictions.

Chopped strand mat can exhibit similar behaviour although the strengths are lower than for continuous fibres and Figure 8.5 shows a tensile stress–strain curve obtained by Allen[11] for a material supplied by Elkalite Ltd. consisting of four layers of chopped strand E-glass mat in high alumina paste, the volume of glass being about 6.7 per cent. It was also shown that the peak values of a cyclic stress–strain curve closely followed the curve in Figure 8.5.

An important structural consideration is the progressive loss of failure strain

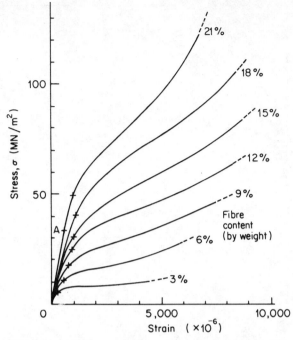

Figure 8.4. Stress–strain curves for high-alumina cement with different volumes of continuous parallel glass-fibre reinforcement-tension along the fibres (Reproduced from Biryukovich, Biryukovich, and Biryukovich,[1] *Glass-fibre-reinforced Cement*, CERA Translation, November 1975, by permission of CIRIA)

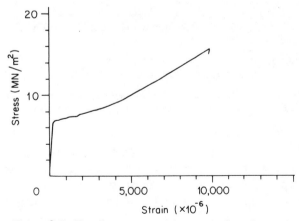

Figure 8.5. Tensile stress–strain curve for chopped strand E-glass mat in high alumina cement ($V_f \simeq 6.7$ per cent) (Reproduced from Allen,[11] *Journal of Composite Materials*, 5, April 1971)

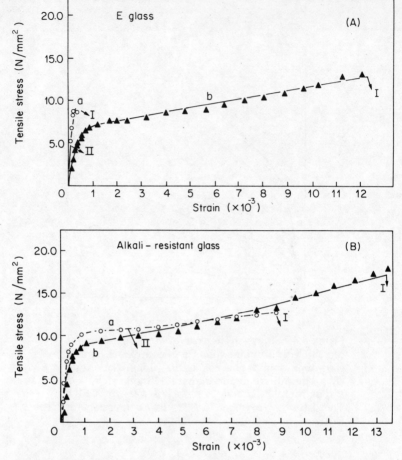

Figure 8.6. The effect of environment on the stress–strain behaviour of glass fibre high-alumina cement

(A) E-glass: (a) water stored, (b) air stored
(B) Alkali-resistant glass: (a) water stored, (b) air stored

(Reproduced from Majumdar,[12] *Fibre-reinforced Cement and Concrete*, 1, 307 (1975) by permission of the publishers, The Construction Press Ltd.)

with time under water storage conditions. Majumdar[12] has tested 34 mm lengths of E-glass and alkali resistant glass at 4 per cent by volume in a 2-D random array in high alumina cement. Figure 8.6 shows the stress–strain curves at one and four years (I and II respectively on Figure 8.6) for the two fibre types under air and water storage.

It can be seen from Figure 8.6 that under wet conditions the composites suffer a serious loss in pseudo-ductility with time which is significantly less pronounced for alkali-resistant glass than for E-glass. For air storage, point II on the figure may be taken as the same as point I and the large strain to failure is maintained for several years.

8.4.2.2 Flexural strength or modulus of rupture

Biryukovich[1] pointed out that the modulus of rupture was higher than the tensile strength, being about 88 MN/m^2 at 9 per cent by weight of parallel continuous fibres, and also that the mode of failure varied with the glass content. For parallel fibres, failure occurred in the tensile zone for 3 per cent and 6 per cent of fibres by weight but at 9 per cent the compression and tension zones failed simultaneously. At glass contents of 12 per cent or more failure took place only in the compression zone.

Composites with 2-D discontinuous fibres were tested by Allen[11] who showed that at 6.7 per cent of fibres by volume the nominal flexural strengths varied from 24.1 MN/m^2 to 37.4 MN/m^2, the position of the neutral axis at failure reaching 0.2 of the section depth from the compression face.

8.4.2.3 Compressive strength

Compressive strengths of more than 70 MN/m^2 have been obtained by Allen[11] although those quoted Biryukovich were less than 50 NM/m^2. This property is particularly sensitive to cement type and w/c ratio.

8.4.2.4 Impact strength

Impact strength has been shown to reduce with time for water storage conditions although alkali-resistant glass is less susceptible than E-glass. Figure 8.7 is reproduced from work by Majumdar.[3]

8.4.2.5 Durability

It was shown by Biryukovich[1] that for 9 per cent by weight of glass fibre in a gypsum-aluminous slag cement the strengths in tension, compression, and bending were virtually unchanged by two years of storage in air in a laboratory.

However, it has been shown conclusively by Majumdar[3] that water storage of composites containing a random array of 2-D short fibres suffer a continuous reduction in the modulus of rupture for at least two years. The loss of flexural strength of the composite containing alkali-resistant fibres is less marked than for E-glass but is nevertheless significant. Figure 8.8 shows the relevant data.

The reduction in impact strengths and moduli of rupture with time do not necessarily imply that the glass has suffered a loss in strength because other factors such as changing bond strength and changing the microstructure of the paste at the fibre interface together with alterations in the composite ductility affect the properties in bending as explained in Sections 4.5 and 8.5.2. Nevertheless the properties of the composite are of overall significance in terms of durability and these have been shown to be time sensitive under certain conditions.

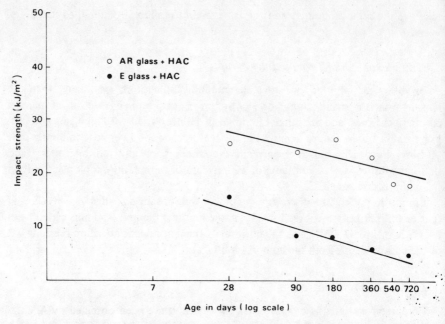

Figure 8.7. Impact strength against log time for E-glass and alkali resistant glass in high alumina cement (Reproduced from Majumdar and Nurse, *Building Research Establishment Current Paper CP 79/74*, August 1974, by permission of The Controller, HMSO. Crown copyright reserved)

Chan and Patterson[13] have carried out durability trials in which they found no significant deterioration in tensile strength under natural weathering conditions or in laboratory air during a period of 1½ years. However, artifical weathering cycles which included periods at 30 °C did show some decrease in strength during a 5 month period.

An explanation for these differing short-term durability results may be that durability is dependent on the value of fibres used and their orientation. For instance, in the case of Biryukovich[1] with 9 per cent by weight of glass fibre any fall in strength might not show up for several years.

Allen[14] has confirmed the conclusions of Biryukovich[1] that specimens kept in laboratory air at normal indoor temperatures show no deterioration with time. However, the important conclusion from this work[14] is that the tensile strength of specimens kept out of doors in a temperate climate reduced from 15 to 9.3 MN/m^2 in 16 months and, equally important, the failure strain reduced from 12,400 $\times$ 10^{-6} to $980^{-6} \times 10^{-6}$. Immersion in water at 35 °C caused even more drastic reductions so that the composite was not much stronger or more ductile than the original matrix. Tests have also shown that immersion in cold water is almost as severe in its effects as immersion in warm water.

These conclusions effectively imply that E-glass fibre in high alumina cement is not a satisfactory material for use in external applications.

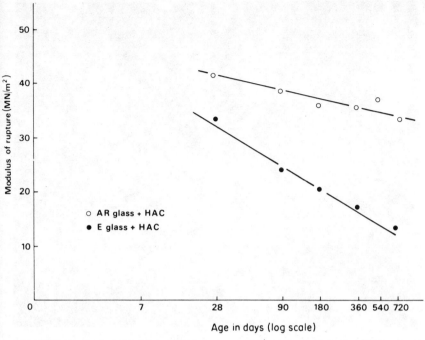

Figure 8.8. Modulus of rupture against log time for water stored E-glass and alkali-resistant glass in high alumina cement (Reproduced from Majumdar and Nurse, *Building Research Establishment Current Paper CP 79/74*, August 1974, by permission of The Controller, HMSO. Crown copyright reserved)

8.5 GLASS FIBRES IN ORDINARY PORTLAND CEMENT COMPOSITES

8.5.1 General

The main emphasis on research and practical applications of glass fibres in cement paste has been concerned with the behaviour of the fibres in ordinary Portland cement. The physical performance of the fibres in this matrix, and, as a result, the performance of the composite is critically dependent on the chemical and physical microstructure of the matrix and of the fibres in the interfacial region where the fibres and matrix make contact. This region is important not only on the external surface of the fibre bundle but also within the bundle itself.[15]

An enlarged view of the critical region is shown in Figure 8.9.

The interfacial region is difficult to study in terms of meaningful physical parameters and therefore many early experimental programmes were carried out on properties of the composite, and the probable changes taking place in the strength of the fibres were deduced from changes in composite properties. Much of the published data regarding the durability of different types of glass fibre has relied on this deductive process but because the relative contributions of changes in the microstructure of the matrix and changes in fibre strength towards changing

Figure 8.9. Enlarged view of the interfacial region between fibres and cement (Reproduced by permission of the Director, Building Research Establishment)

composite properties were not quantified, conclusions regarding the durability of the glass fibres themselves in a Portland cement matrix should be treated with caution. Also, after much of the early work was published, it was shown that composite tensile strengths calculated from flexural tests are dependent on the composite strain capacity, independently of the tensile strength (Section 4.5) which could have given a misleading impression regarding fibre durability.

It follows that the durability of glass fibres must be judged from strength tests on the fibres alone after they have been removed from the matrix. This is a very difficult experimental process and reliable data on the effects of alkali chemical attack on the strength of the different types of glass fibres by the matrix are therefore limited.

Another area which may affect the performance of the fibre in a cement matrix and about which there is little quantitative data is the effects of different types of size or surface coatings which are applied to the fibres immediately after drawing from the melt. Direct comparisons between different glasses cannot therefore be made unless details of type and thickness of surface coating are available.

8.5.2 Microstructure of the glass fibre—cement interface

The interface between alkali-resistant glass fibres and cement paste has been studied in great detail by Stucke and Majumdar[15] and Figures 8.10 to 8.13 are

Figure 8.10. Pull out grooves showing fibre matrix interfaces in 90 day dry air stored glass reinforced cement (Reproduced by permission of the Director, Building Research Establishment)

similar to those in Reference 15. The critical factor which controls the properties of the composite appears to be the density of the interfacial region and this density depends on time and storage conditions.

Figure 8.10 shows the pull out grooves in 90 day, dry air (40 per cent R.H.) stored material in which the interface consists of fine whisker-like crystals forming an interpenetrating mat in the voids between partially hydrated cement grains.

Interfaces after 5 years dry air storage are very similar to those at 90 days. The contact area between fibre and cement is relatively low for these conditions leading to a low frictional shear strength and a relatively compliant matrix at the point where a fibre is bent across a crack. These factors result in the composite having a high toughness due to the extensive fibre pull out which can be seen in Figure 8.11.

In contrast to Figure 8.10, the fibre—matrix interfaces shown in Figure 8.12 for 5 year water stored material are very dense with a residual porosity of almost zero and a high contact area leading to a greatly increased frictional bond and a hard stiff material at the point where the fibres bend across a crack.

These combined effects lead to the type of fracture surface shown in Figure 8.13 with little fibre pull out and with the majority of the fibre fractures occuring near the crack in the matrix. The result is a material with greatly reduced toughness or resistance to impact loads.

The appearance of the interface in 5 year naturally weathered material varies

Figure 8.11. Fracture surface of glass reinforced cement stored for 5 years in dry air at 20 °C (Reproduced by permission of the Director, Building Research Establishment)

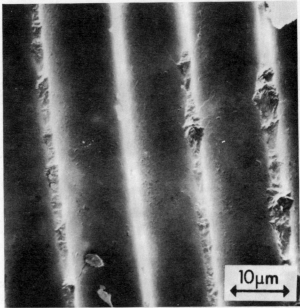

Figure 8.12. Fibre-matrix interfaces in 5 year water stored glass reinforced cement (Reproduced by permission of the Director, Building Research Establishment)

Figure 8.13. Fracture surface of glass reinforced cement stored for 5 years in water at 20 °C (Reproduced by permission of the Director, Building Research Establishment)

from porous to fully dense and as a result, the composite exhibits properties with some of the characteristics of the wet and of the dry storage conditions.

8.5.3 Corrosion of E-glass fibres by cement paste

No information has been found regarding the tensile strengths of E-glass fibres removed from Portland cement composites.

However, stereoscan photographs have been published[16] showing the surface condition of the fibres extracted from cement composites after one year indoor storage. Substantial surface-pitting is visible which could lead to strength loss because the strength of glass fibres is dependent on the size and population of the surface flaws.

Majumdar[17] and Cohen et al.[18] have measured the change in strength of E-glass fibres kept in a 'cement extract' solution at various temperatures but due to the difference in availability of alkali solution at the fibre interface in hardened cement paste when compared with a concentrated solution immediately in contact with the fibre surface there can be no direct correlation with corrosion in glass-fibre–cement composites.

The basic chemistry of the alkali attack has been described by Larner et al.[19] and there is no doubt from this work that E-glass fibres corrode much more quickly than alkali-resistant fibres in a cement extract solution.

8.5.4 Corrosion of alkali resistant glass fibres by cement paste

Alkali resistant glass fibres have been removed from test boards made from ordinary Portland cement by two authorities,[18,20] and then tested for tensile strength.

Cohen and Diamond[18] used air stored boards at 22 °C and 50 per cent R.H. and found that the fibre strength reduced between mixing and 1 day but that no change in strength was observed after a further 500 days of exposure.

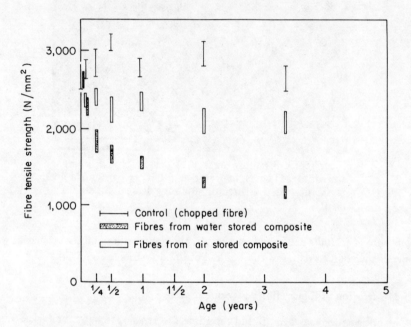

Figure 8.14. Tensile strength of alkali-resistant glass fibres extracted from cement composites (The bars represent 90 per cent confidence limits.) (Reproduced from Majumdar, West, and Larner,[21] *Journal of Materials Science*, 12, 927–936 (1977), by permission of Chapman and Hall Ltd.)

Majumdar[20] carried out strength tests on fibres after composites had been stored either in air or in water for periods up to 1200 days. Figure 8.14 shows that the strength reduction in fibres removed from water stored composites is considerably greater than that from air stored material but there are indications from tests with cement extract solutions that the strength may not fall much below 1000 MN/m^2 for periods up to 5 years.

The level of residual strength observed in Majumdar's work was about 30 per cent higher than that obtained by Cohen and Diamond.

8.5.5. Direct tensile properties of composite containing glass fibres

8.5.5.1 Effect of length and volume of alkali-resistant fibres on strength and stress–strain behaviour [6]

The effect of length and volume of fibre on the properties of spray–suction composites has been investigated by Ali et al.[6] Fibre lengths between 10 mm and 40 mm were used at volumes between 2 per cent and 8 per cent and the comparison of properties was made at 28 days. The importance of the microstructural features of the interface between fibre and matrix has been stressed in Section 8.5.2 and it follows that the conclusions drawn from the 28 day results may not be applicable at later ages under different exposure conditions.

Figure 8.15 shows the effect of fibre volume on tensile strength for four different fibre lengths, the higher strengths of the composites with longer fibres being attributed to the increased length efficiency factors (Section 3.4.7).

The porosity of the composite increases as the fibre volume increases (Figure 8.16) and this effect may be the cause of the reduction in tensile strength of some of the composites at fibre volumes greater than 6 per cent.

Figures 8.17 and 8.18 are self explanatory and show the type of stress–strain curves which have been described theoretically in Section 3.4. Suitable values of material parameters applicable to these composites are given in Table 8.2.

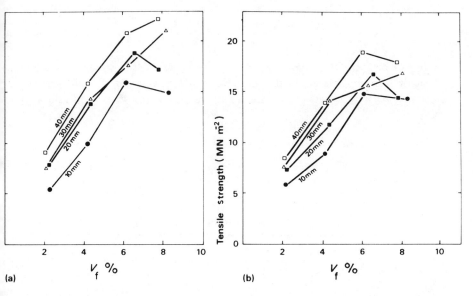

Figure 8.15. Relation between fibre volume and tensile strength of glass reinforced cement at 28 days for different fibre lengths (a) Stored in air (b) Stored in water (Reproduced from Ali, Majumdar, and Singh,[6] *Building Research Establishment Current Paper CP 94/75*, October 1975, by permission of The Controller, HMSO. Crown copyright reserved)

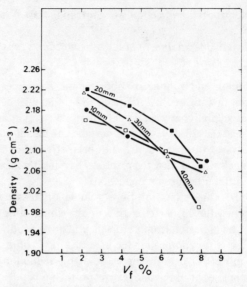

Figure 8.16. Relation between fibre volume and density of glass reinforced cement composites at 28 days for different fibre lengths (Reproduced from Ali, Majumdar, and Singh,[6] *Building Research Establishment Current Paper CP 94/75*, October 1975, by permission of The Controller, HMSO. Crown copyright reserved)

It can be seen from these figures that the largest tensile strengths and strain capacities can be achieved, in the short term, by high volumes of the longest fibres and that differences between air and water storage are not very significant at this age.

Table 8.2. Material parameters applicable to the composites shown in Figure 8.15 to 8.18 (Reproduced from Ali, Majumdar, and Singh, *Building Research Establishment Current Paper CP 94/75*, October 1975, by permission of the Controller, HMSO, Crown Copyright reserved)

Parameter	Typical value
Fibre–cement bond (τ)	3 MN/m^2
Modulus of cement paste (E_m)	26 GN/m^2
Modulus of composite (E_c)	30 GN/m^2
Modulus of fibre (E_f)	76 GN/m^2
Effective area of glass fibre strand (A_f)	0.027 mm^2
Perimeter of glass fibre strand (P_f)	1.42 mm

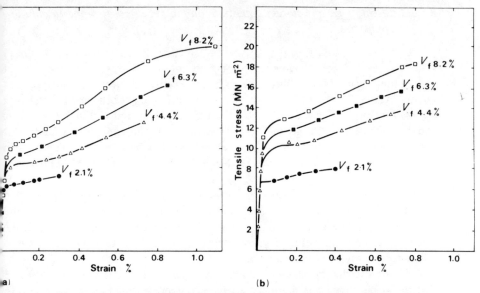

Figure 8.17. Tensile stress–strain curves of glass reinforced cement composites containing 30 mm long fibres at 28 days (a) Stored in air (b) Stored in water (Reproduced from Ali, Majumdar, and Singh,[6] *Building Research Establishment Current Paper CP 94/75*, October 1975, by permission of The Controller, HMSO. Crown copyright reserved)

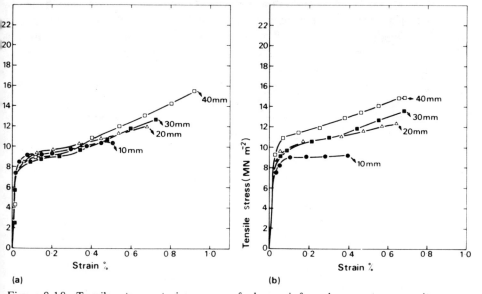

Figure 8.18. Tensile stress–strain curves of glass reinforced cement composites containing 4 per cent of fibres by volume with different fibre lengths at 28 days (a) Stored in air (b) Stored in water (Reproduced from Ali, Majumdar, and Singh,[6] *Building Research Establishment Current Paper CP 94/75*, October 1975, by permission of The Controller, HMSO. Crown copyright reserved)

8.5.5.2 Effect of age and storage conditions on the direct tensile properties of the composite

The effect of storage in dry air, natural weather, and water storage has been studied in detail[8] for spray-dewatered boards containing 5 per cent by weight of 34 mm long alkali resistant fibres in an ordinary Portland cement matrix for periods up to 5 years. The tensile strength of dry air-stored samples showed little change during the five year period but the wet stored specimens reduced in strength from about 15 MN/m^2 to 10 MN/m^2 during the first year and appeared to stabilize at about this value. On the other hand, the naturally weathered samples appeared to be continuing to reduce in strength to below 10 MN/m^2 at the end of the five-year period.

Of perhaps greater significance than the absolute strengths is the change in strain to failure for the different storage condition. Figure 8.19 shows that the failure strain for both the water stored and naturally weathered composites reached the matrix cracking strain before 5 years whereas little change occurred in the strain to failure for the dry air stored material.

Toughness is closely related to failure strain and therefore it would be expected that the water stored and naturally weathered materials would become brittle with time and would loose a higher proportion of impact and flexural strengths than of direct tensile strength (Sections 4.5 and 8.5.2).

The measurements of tensile properties up to 5 years have been used to estimate the properties at 20 years for a variety of exposure conditions.[8] In the case of naturally weathered material these estimates depend greatly on the method used to extrapolate the data, and all the extrapolation methods rely on the assumption that the strength–time curve is continuous without step changes or sudden changes in the fracture mechanism. It has already been shown (Section 6.6) that such changes can occur in fibre concretes, particularly in the case of the durability of steel fibre concretes, and therefore the predicted 20-year values must be treated with some caution.

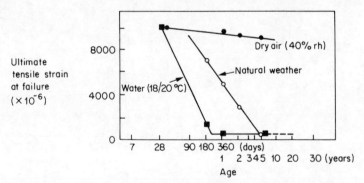

Figure 8.19. Strain to failure in tension for glass reinforced cement at various ages in dry air, natural weather and water storage (Reproduced from *Building Research Establishment Current Paper CP 38/76*, June 1976, by permission of The Controller, HMSO. Crown copyright reserved)

Table 8.3 has been obtained using data from the Building Research Establishment[8] and shows that although the Young's modulus is not greatly changed with time for natural weathering conditions, the tensile strength using the most pessimistic assumptions may reduce from 17 MN/m^2 to 4 MN/m^2 which is approximately the strength of the matrix.

The overall durability situation is complicated by the possible use in the future of matrices containing pulverized fuel ash or supersulphated cements which can give radically different durability curves.

Also, commercial GRC contains about 30 per cent of sand to reduce shrinkage and durability data on these materials is several years behind that on plain cement composites.

8.5.6 Flexural properties of spray-dewatered composites containing alkali-resistant glass fibres

8.5.6.1 Strains in bending

It has been shown in Chapter 4 that a critical factor affecting the calculated nominal flexural strength of a fibre—cement composite is the position of the

Table 8.3. Measured and predicted direct tensile properties of spray-dewatered glass reinforced ordinary Portland cement (5 per cent glass fibre) B.R.E. data[8] (Reproduced from *Building Research Establishment Current Paper CP 38/76*, June 1976, by permission of The Controller, HMSO, Crown copyright reserved)

Age	Storage condition	Ultimate tensile strength MN/m^2	Young's modulus GN/m^2
28 days	Range for air and water storage	14–17	20–25
1 year	Air[a]	14–16	20–25
	Water[b]	9–12	28–34
	Weathering	11–14	20–25
5 years	Air[a]	13–15	20–25
	Water[b]	9–12	28–34
	Weathering	7–8	25–32
20 years (Estimated)	Air[a]	12–15	20–25
	Water[b]	8–11	28–34
	Weathering[c]	log 4–6	
		exp 7–10	25–32

[a] At 40 per cent relative humidity and 20 °C.
[b] At 18–20 °C.
[c] log = Logarithmic extrapolation of data assumes continual decay throughout the 20 years.
exp = Exponential extrapolation assumes equilibrium values achieved within 20 years.

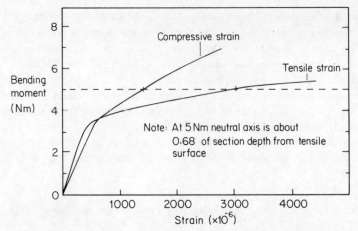

Figure 8.20. Flexure test on an O.P.C. – alkali-resistant glass-fibre composite showing the difference in measured strains between compressive and tensile surfaces (Reproduced from Allen, *Report 55*, September 1975, by permission of CIRIA)

neutral axis at failure which, in turn, is defined by the relative magnitudes of compressive and tensile strains on the opposite surfaces of the beam. Because the flexural strength, or modulus of rupture, is calculated from the erroneous assumption that the neutral axis remains at the mid-point of the section up to failure the values quoted for flexural strength are fictitious and do not represent stresses present in the specimen or the tensile strength of the material. Nevertheless, the strengths can be used to calculate the strengths of components of similar thickness subjected to bending stresses.

This point is illustrated by measurements of compressive and tensile strains made by Allen[14] on the surface of a beam produced from a spray-suction composite. (Figure 8.20) The nominal flexural strength was between 20 and 28 MN/m^2 based on the assumption that the neutral axis remained at mid-depth (0.5D) but although this was the case at a bending moment of 3.5 Nm, the neutral axis reached 0.68D from the tensile face at 5 Nm and continued to move nearer to the compression surface thereafter.

8.5.6.2 Effect of length and volume of alkali-resistant fibres on the apparent modulus of rupture

The results shown in Figure 8.21 were obtained by Ali *et al.*[6] and are for the same composites described in Section 8.5.5.1. Similar conclusions apply as for the tensile strength.

8.5.6.3 Effect of age and storage conditions on the flexural properties of the composite[8]

Similar trends are shown for air storage, water storage, and natural weathering as have already been described for direct tensile strength in Section 8.5.5.2. Natural

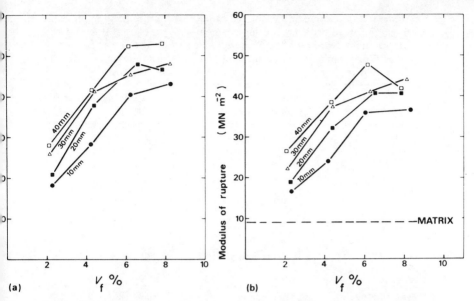

Figure 8.21. Relation between fibre volume and modulus of rupture of glass reinforced cement at 28 days for different fibre lengths. (a) Stored in air (b) Stored in water (reproduced from Ali, Majumdar, and Singh,[6] *Building Research Establishment Current Paper CP 94/75*, October 1975, by permission of The Controller, HMSO. Crown copyright reserved)

weathering appears to produce about the same reduction in flexural strength at 5 years as water storage at 20 °C but the water stored strength appears to have reached comparative stability while the naturally weathered strengths may be continuing to decrease with time.

Table 8.4 shows the effect of time on the flexural strength of a particular composite.

Some of the difficulties of predicting strength for periods up to 50 years are emphasized by Figure 8.22. In Figure 8.22, the relatively large scatter of results combined with the small changes in average strength observed make extrapolation very difficult over a time span of 10 times that covered by the experimental data.

8.5.7 Impact properties

Quantitative values for impact strength are not of great interest for design purposes because they are not fundamental material parameters. The measurements are highly dependent on the techniques used to measure the property and, in particular, variations in the mass and velocity of the impacting device can alter the estimate of 'toughness' by more than 100 per cent. However, Izod tests have often been used for glass-reinforced cement and comparative values for the effects of age, fibre volume, and fibre length are valuable.

The impact strengths of air and water stored spray-suction composites at 28 days are shown in Figure 8.23.

Table 8.4. Measured and predicted moduli of rupture of spray-dewatered glass reinforced ordinary Portland cement (5% glass fibre) B.R.E. data (Reproduced from *Building Research Establishment Current Paper CP 38/76*, June 1976, by permission of The Controller, HMSO, Crown Copyright reserved)

Age	Storage condition	Modulus of rupture MN/m^2
28 days	Range for air and water storage	35–50
1 year	Air	35–40
	Water[b]	22–25
	Weathering	30–36
5 years	Air[a]	30–35
	Water[b]	21–25
	Weathering	21–23
20 years (Estimated)	Air	26–34
	Water	20–25
	Weathering[c]	log 12–15
		exp 21–23

[a] At 40 per cent relative humidity and 20 °C
[b] At 18–20 °C
[c] log = Logarithmic extrapolation of data – assumes continual decay throughout 20 years
  exp = Experimental extrapolation equilibrium values achieved within 20 years

Figure 8.23 shows that water stored samples, even at 28 days, are weaker than air stored specimens and it has been found that this difference increases with time. Air stored material at 40 per cent relative humidity is likely to maintain its toughness for periods of up to 20 years but both water stored and naturally weathered material approach the impact strength of the matrix (about 2 to 5 KJ/m^2) within the first few years.

The reduction in impact strength is attributed to increased fibre breakage and reduced fibre pull out as the density of the matrix increases with time (Section 8.5.2).

Fortunately the impact strength is relatively high in the first month or two, which is the important time for handling stresses, but the increasing brittleness requires a certain amount of caution when semi-structural applications are being considered.

8.5.8 Fatigue strength

Fatigue tests have been carried out for spray de-watered composites in direct tension[8] and in flexure[7] and representative results are shown in Figures 8.24 and 8.25.

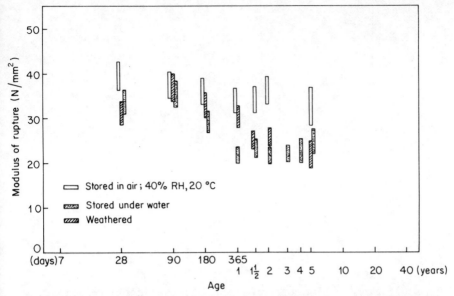

Figure 8.22. Modulus of Rupture against log time for cement composites made with alkali resistant glass stored in various environments (Reproduced from Majumdar, *Fibre-reinforced Cement and Concrete,* **1**, 306 (1975) by permission of the publishers, The Construction Press Ltd.)

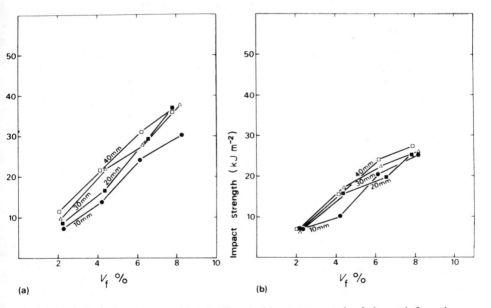

Figure 8.23. Relation between fibre volume and impact strength of glass reinforced cement at 28 days (a) Stored in air (b) Stored in water (Reproduced from Ali, Majumdar, and Singh,[6] *Building Research Establishment Current Paper CP 94/75,* October 1975, by permission of The Controller, HMSO. Crown copyright reserved)

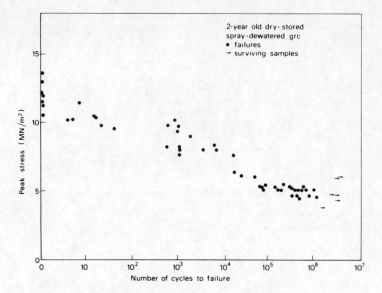

Figure 8.24. Direct tensile fatigue of glass reinforced cement in air (Reproduced from *Building Research Establishment Current Paper CP 38/76*, June 1976, by permission of The Controller, HMSO. Crown copyright reserved)

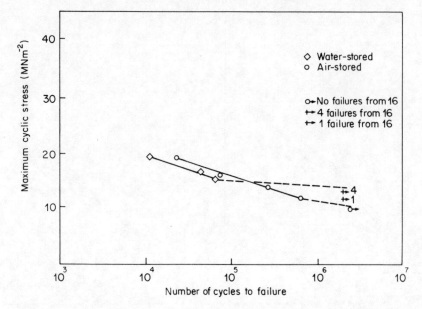

Figure 8.25. Flexural fatigue of glass reinforced cement in air and water (Reproduced from Hibbert and Grimer,[7] *Building Research Establishment Current Paper CP 12/76*, January 1976, by permission of The Controller, HMSO. Crown copyright reserved)

Fatigue lives of more than 10^6 cycles are possible at tensile stresses at 4.5 MN/m² or at bending stresses of 10 MN/m² in both prolonged wet or dry storage conditions.

8.5.9 Compressive strength[8]

The across-plane compressive strength is equal to that of the matrix, which is about 100 MN/m² for cement paste, but at right angles to this direction the strength may be somewhat lower at about 65 MN/m².

8.5.10 Pre-mixed glass fibre cement

The properties of pre-mixed glass fibre cement have not been as extensively researched as those of the spray dewatered composite but the static strengths are likely to be lower for a given percentage of glass than those quoted for the sprayed material because the orientation of the fibres will probably be less efficient than 2-D random. Figure 8.26 shows the measured moduli of rupture and impact strengths for glass contents up to 15 per cent by weight and if these results are

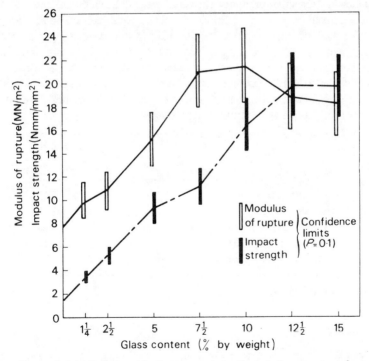

Figure 8.26. Relationship between strength and glass content for pre-mixed glass-reinforced cement at 28 days (O.P.C. and alkali resistant glass) (Reproduced from Hills,[4] *Building Research Establishment Current Paper CP 65/75*, July 1975, by permission of The Controller, HMSO. Crown copyright reserved)

compared with Figures 8.21 and 8.23 for the sprayed material at the same age it can be seen that the strengths of pre-mix are about 50 per cent lower.

8.5.11 Combinations of glass and polypropylene fibres

It may be advantageous to combine different fibres in the same matrix to more fully utilize the benefits from each fibre type. Polypropylene and glass fibres have been mixed in various concentrations by Walton et al.[5] and it is probable that initial stiffness can be provided by the glass while long term toughness and durability will be assisted by the use of polypropylene.

8.6 GLASS FIBRES IN CONCRETE

8.6.1 General

Although glass fibres have been used in overlays to concrete pavements (Section 11.3) there is relatively little published work on the physical properties of glass fibre concretes. This is because the research effort has been concentrated into the development of thin sheet products which require fine grained matrices.

However, alkali resistant glass fibres in concrete have been studied by Marsh and Clark[22] for the limited condition of testing after 14 days and curing in air at 50 per cent R.H. and 23 °C. Fibre lengths between 12 mm and 50 mm were used at volumes between 0.5 per cent and 2.5 per cent. Various mix proportions were examined using 10 mm maximum sized river gravel aggregate, water reducing admixtures, and about 5 per cent of entrained air. The aggregates were saturated surface dry (SSD) and a typical mix is shown below:

	Batch weights (kg/m^3)
Cement	502
Water	251
Coarse aggregate (SSD)	340
Fine aggregate (SSD)	1028
Glass fibres (1 per cent by volume)	27

Compaction was by external vibration.

8.6.2 Workability

The effect of fibre volume on the slump of the typical mix is shown in Figure 8.27 from which it can be seen that, as with other types of fibre, the slump decreases rapidly with increasing volume reaching zero for all fibre lengths at 2.5 per cent by volume. The 12 mm fibres gave a higher slump that the longer fibres but all mixes were consolidated without great difficulty.

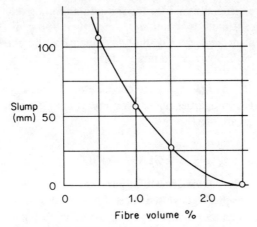

Figure 8.27. Effect of fibre content on slump (Reproduced from Marsh and Clarke,[22] *Fibre-reinforced Concrete, Publication SP 44*, 1974, p. 263, by permission of American Concrete Institute)

8.6.3 Flexural strength

The flexural strength was obtained from third point loading on 76 mm x 102 mm beams with a 305 mm span. Figure 8.28 shows that the ultimate strength increases with increasing fibre volume up to about 2 per cent of fibres.

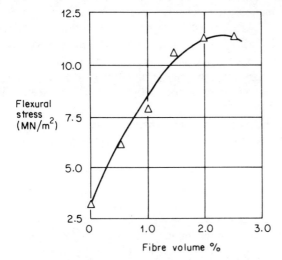

Figure 8.28. Effect of fibre content on ultimate flexural strength (Reproduced from Marsh and Clarke,[22] *Fibre-reinforced Concrete, Publication SP 44*, 1974, p. 262, by permission of American Concrete Institute)

The 12 mm fibres generally gave the lowest strength for a given fibre volume but the effect of fibre length was not very consistent, possibly because of the interacting effects of decreasing workability and increasing bond area. However the 38 mm fibres were probably most promising at this age in terms of ultimate flexural strength.

It is inevitable that before glass fibre concrete can be used with confidence in practice, the long term properties under a variety of curing conditions will have to be studied in much greater detail.

8.7 GLASS WOOL

Glass wool is produced for insulating purposes by blowing or spinning from molten glass and Krenchel[23] has used these fibres in the form of a paper made from 5 μm fibres, 12 mm long at a volume fraction of about 6 per cent. The cement paste is scraped through the paper and the material is built up from laminates with a 2-D orientation of fibres. The specific fibre surface is about 50 mm^2/mm^3 and this results in invisible microcracks forming on the tensile face of flexure specimens which can have moduli of rupture in excess of 25 MN/m^2. However, the glass used in the tests was not alkali-resistant and further development is required before composites with long term durability can be guaranteed.

8.8 GLASS FIBRES IN GYPSUM PLASTER[3,24,25]

Gypsum plaster is a brittle matrix which was originally used in conjunction with E-glass fibres at the Building Research Establishment[24] in order that the development of techniques for the manufacture of glass-reinforced cement components would not be delayed by the initial lack of availability of bulk quantities of alkali resistant glass fibres. It was found during tests on composites produced by mixing and by the spray-suction process for sheet materials (Section 8.3.7) that glass reinforced gypsum is a useful material in its own right even although the 50 per cent loss in strength when wetted restricts it to interior uses where it is protected from the weather.

Two types of matrix have been used, the α-hemihydrate which has a lower water requirement for a workable mix and hence gives higher density and strength and the β-hemihydrate or 'plaster of Paris'.

8.8.1 Properties[3,2,25]

The two most valuable properties of glass-reinforced gypsum are impact strength and fire resistance. The impact strength is about double that of glass reinforced cement having the same glass content, probably due to the weaker fibre—matrix bonding which allows more energy absorption by fibre pull out. The fire resistance is attributed, in part to the 18 per cent of water combined in the gypsum crystal and partly due to the reinforcing effect of the fibres which prevent the material shattering in a fire.

Table 8.5. Properties of gypsum plasters (Reproduced from Majumder and Nurse,[3] *Building Research Establishment Current Paper CP 79/74*, August 1974, by permission of The Controller, HMSO, Crown Copyright reserved)

	Plaster of Paris	α-Hemihydrate
SO_3 %	38.6	39.0
CaO %	52.5	52.7
Loss on ignition at 290 °C %	6.3	6.1
Transverse strength MN/m^2 (BS 1191–1967)	7.2	12.1

The properties of the plaster matrices are shown in Table 8.5 in which the α-hemihydrate contained a retarder and 0.1 per cent Keratin was added to the plaster of Paris.

The main strength properties of the composites vary with fibre volume, fibre length and fibre orientation in a similar way to those previously described for glass reinforced cement and tensile strengths of up to 18 MN/m^2 are possible for spray-suction composites.

The effect of glass content for 50 mm long fibres in a spray-suction composite is shown in Figure 8.29.

Both plasters show an optimum addition of fibre after which the increasing difficulty of compaction more than offsets the increased reinforcing effect.

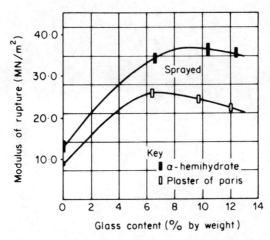

Figure 8.29. Modulus of rupture and glass content of glass reinforced gypsum (Reproduced from Majumdar and Nurse,[3] *Building Research Establishment Current Paper CP 79/74*, August 1974, by permission of The Controller, HMSO. Crown copyright reserved)

8.9 ASSESSMENT OF GLASS-FIBRE QUANTITY AND DISTRIBUTION IN OPAQUE COMPOSITES[26]

It is important to be able to determine the quantity and distribution of glass fibre in laboratory specimens and in commercial products and Hibbert[26] has developed a technique which utilizes the transmission of light along the fibres by internal reflection when the fibres are embedded in an opaque matrix such as

Figure 8.30. Section of an aligned glass-fibre-reinforced cement composite taken perpendicular to fibre direction (Reproduced from Hibbert, *Journal of Materials Science, Letters*, **9**, 512–514 (1974) by permission of Chapman and Hall Ltd.)

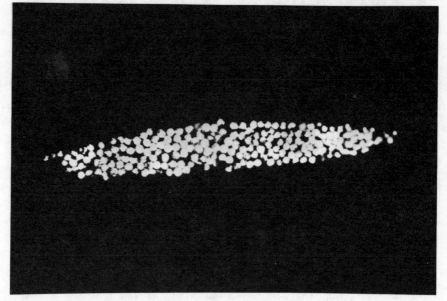

Figure 8.31. Photograph of a single strand in the composite shown in Figure 8.30 (Reproduced from Hibbert, *Journal of Materials Science, Letters*, **9**, 512–514 (1974) by permission of Chapman and Hall Ltd.)

cement or plaster. In this technique, sections of the composite of a thickness suitably less than the fibre length (say 5 mm to 10 mm) are cut using a water-fed, diamond impregnated, cutting wheel. Illumination is provided on one surface and the other surface is photographed, typical examples being shown in Figures 8.30 and 8.31.

In these Figures the distribution and orientation of the strands is clearly visible and Figure 8.31 shows the individual filaments in one of the strands.

The method also lends itself to computerized counting and analysis of fibre volume and distribution.

REFERENCES

1. Biryukovich K. L., Biryukovich, Yu. L, and Biryukovich, D. L., *Glass-fibre-reinforced Cement*, Published by Budivel'nik, Kiev, 1964, CERA Translation, No. 12, Nov. 1965.
2. Majumdar, A. J., 'The role of the interface in glass-fibre-reinforced cement,' *B.R.E. Current Paper*, CP 57/74, May 1974.
3. Majumdar, A. J., and Nurse, R. W., 'Glass-fibre-reinforced cement,' *Building Research Establishment Current Paper*, CP 79/74, August 1974.
4. Hills, D. L., 'Premixed glass-fibre-reinforced cement,' *Building Research Establishment Current Paper*, CP 65/75, July 1975.
5. Walton, P. L., and Majumdar, A. J., 'Cement based composites with mixtures of different types of fibres,' *Building Research Establishment Current Paper*, CP 80/75, September 1975.
6. Ali, M. A., Majumdar, A. J., and Singh, B., 'Properties of glass-fibre cement — the effect of fibre length and content,' *Building Research Establishment Current Paper*, CP 94/75, October 1975.
7. Hibbert, A. P., and Grimer, F. J., 'Flexural fatigue of glass-fibre reinforced cement,' *Building Research Establishment Current Paper*, CP 12/76, January 1976.
8. 'A study of the properties of Cem-FIL/OPC Composites,' *Building Research Establishment Current Paper*, CP 38/76, June 1976.
9. Krenchel, H., 'Fibre-spacing and specific fibre surface,' *Fibre-reinforced Cement and Concrete*, RILEM Symposium, pp. 69–79 (1975).
9a. Oakley, D. R., and Proctor, B., 'Tensile stress–strain behaviour of glass-reinforced cement composites', *Fibre-reinforced Cement and Concrete*, RILEM Symposium 1975, pp. 347–359.
10. Larner, L. J., Speakman, K., and Majumdar, A. J., 'Chemical interaction between glass-fibres and cement,' *Journal of Non-crystalline Solids*, 20, 43–74 (1976).
11. Allen, H. G., 'Stiffness and strength of two glass-fibre reinforced cement laminates,' *Journal Composite Materials*, 5, 194–207 April, (1971).
12. Majumdar, A. J., 'Properties of fibre–cement composites,' *Fibre reinforced Cement and Concrete*, RILEM Symposium, 1975, pp. 279–313.
13. Chan, H. C., and Patterson, W. A., 'Effects of ageing and weathering on the tensile strength of glass-fibre-reinforced high alumina cement,' *Journal of Materials Science*, 6, 342–346 (1971).
14. Allen, H. G., 'Glass-fibre-reinforced cement — strength and stiffness,' *CIRIA Report*, 55.
15. Stucke, M. S., and Majumdar, A. J., 'Microstructure of glass fibre-reinforced cement composites,' *Journal of Materials Science*, 11, 1019–1030 (1976).

16. Jaras, A. C., and Litherland, K. L., 'Microstructural features in glass-fibre-reinforced cement composites,' *Fibre-reinforced Cement and Concrete*, RILEM Symposium, 1975, Construction Press Ltd., pp. 327–334.
17. Majumdar, A. J., 'The role of the interface in glass-fibre-reinforced cement,' *Composites*, January 1975, pp. 7–16.
18. Cohen, E. B., and Diamond, S., 'Validity of flexural strength reduction as an indication of alkali attack on glass in fibre-reinforced cement composites,' *Fibre-reinforced Cement and Concrete*, RILEM Symposium, 1975, pp. 315–325.
19. Larner, L. J., Speakman, K., and Majumdar, A. J., 'Chemical interactions between glass fibres and cement,' *Journal of Non-crystalline Solids*, **20**, pp. 43–74 (1976).
20. Majumdar, A. J., 'Properties of fibre cement composites,' *Fibre-reinforced Cement and Concrete*, RILEM Symposium, 1975, pp. 279–313.
21. Majumdar, A. J., West, J. M., and Larner, L. J., 'Properties of glass fibres in cement environment,' *Journal Materials Science*, **12**, 927–936 (1977).
22. Marsh, H. N., and Clarke, L. L., 'Glass fibres in concrete,' ACI Publication S.P. 44, *Fibre-reinforced Concrete*, 1974, pp. 247–264.
23. Krenchel, H., and Hejgaard, O., 'Can asbestos be completely replaced one day?' *Fibre-reinforced Cement and Concrete*, RILEM Symposium 1975, pp. 335–346.
24. Ryder, J. F., 'Glass-fibre reinforced gypsum plaster. Prospects for fibre-reinforced construction materials,' *Proceedings of an International Building Exhibition Conference*, London, November, 1971, Published by Building Research Establishment, 1972, pp. 69–89.
25. Laws, V., Lawrence, P., and Nurse, R. W., 'Reinforcement of brittle matrices by glass fibres,' *Journal Physics D. Applied Physics*, **6**, 1973.
26. Hibbert, A. P., 'A method for assessing the quantity and distribution of glass fibre in an opaque matrix.' *Journal of Materials Science Letters*, **9**, 512–514 (1974).

Chapter 9
Asbestos Cement

9.1 ASBESTOS FIBRES

9.1.1 Mineralogy and production

Asbestos is a general name for several varieties of naturally occurring crystalline fibrous silicate minerals which possess a rather unique range of physical and chemical properties. The two main groups are the serpentines and the amphiboles and generally both types have developed as cross-fibre seams or veins in the host rocks. The width of the seams determines the fibre length which is commonly in the range 0.8 to 19 mm,[1] and the fibres exist in extremely tight packed parallel formations. Certain types of asbestos occur in fibrous masses with randomly oriented blocks of fibres up to 25 mm in length and it is not uncommon to find fibres up to 100 mm long.

By far the most abundant mineral is chrysotile ($3MgO.2SiO_2.2H_2O$) or white asbestos and this is the sole member of the serpentine group. Figure 9.1 shows an enlarged view of chrysotile fibres.

Chrysotile constitutes more than 90 per cent of the world asbestos reserves and is used to a large extent in the manufacture of asbestos cement. The fibre is white and silky with a minimum diameter of about 0.01 μm and electron microscope work suggests that even the finest crysotile fibres are hollow which may partly account for their affinity for cement. Also, the fibres resist well the severe mechanical attrition received during processing possibly due to their high strength and flexibility.

Very high values (> 3000 MN/m^2) are quoted in the literature for the tensile strength of chrysotile but the strengths of the fibre bundles in the composite are much lower. For example Klos[2] considers 560–750 MN/m^2 to be a practical range whereas work at the Building Research Establishment[3] has measured 300–1800 MN/m^2 for fibre bundles. Fibre strengths of this order are easily obtained with several fibres other than asbestos and this consideration is largely responsible for the current interest in asbestos substitutes for cement reinforcement. The chemical resistance offered by chrysotile asbestos, particularly to strongly alkaline conditions is considered to be excellent and this is reflected in the durability which

Figure 9.1. Chrysotile asbestos fibres as seen under the scanning electron microscope (Supplied by P. M. Bills, University of Surrey)

asbestos cement products generally enjoy. The fibres are, however, susceptible to strength loss at elevated temperatures and above 400 °C their strengthening power is greatly reduced. As such high temperatures are not normally encountered in buildings, asbestos cement products are considered to be reasonably safe in most applications in which they are currently used.

The strongest type of fibres in the amphibole group is crocidolite or blue asbestos ($Na_2O.Fe_2O_3.3FeO.8SiO_2.H_2O$) and the group also contains amosite, anthophylite, tremolite, and actinolite. Crocidolite is considered to be the most dangerous form of asbestos from the point of view of hazard to health. Several geographical regions of the world have exploitable asbestos mines but the largest deposits are located in Canada. USSR, and Southern Africa. Modern methods of winning and refining asbestos allow economic extraction from rocks with fibre yields as low as 2 per cent. Asbestos fibres produced at the mines require further processing at manufacturing sites to make them suitable for many applications. This involves milling to split the coarser fibres into finer ones. A variety of methods is used, ranging from edgerunners and rod mills to hammer mills and attritors, the choice of particular methods depending on the application.

9.1.2 Hazard to health

It has been known for many years that exposure to asbestos fibres can be injurious to health and the precautions to be taken in industry have been outlined by the Department of Employment.[4]

Asbestosis and bronchial cancers are caused by inhaling asbestos dust over a long period of time and are similar to the occupational diseases brought about by inhaling coal dust, silica dust, and slate dust. Gilson[5] states that all types of commercially available asbestos can cause these diseases.

Mesothelioma is a cancer of the lining of the chest known to be caused by inhalation of asbestos fibres and it is believed that very short exposure to asbestos dust of the right type can lead to mesothelioma.

Stomach cancer is also believed to be caused by ingesting food which contains asbestos and workmen exposed to asbestos show a marked increase in gastro-intestinal cancer, especially stomach cancer.

It is difficult to make a definitive statement regarding the dangers to health of materials such as asbestos cement where the fibres are tightly bound within the matrix but any operation which releases fibres or dust such as sawing or drilling will require precautions to be taken against inhalation or ingestion of the material.

9.2 ASBESTOS CEMENT

Since 1900, the most important example of a fibre cement composite has been asbestos cement. The proportion by weight of asbestos fibre is normally between 9 to 12 per cent for flat or corrugated sheet, 11 to 14 per cent for pressure pipes and 20 to 30 per cent for fire resistant boards[6] and the binder is normally a Portland cement.

According to Klos[2] the important fibre parameters to be checked before inclusion in cement are fibre-length distribution, dry and wet density, dust content, specific surface and filtrability and there are various standard tests to assess these parameters.[7]

9.2.1 Production Technology

Before the asbestos fibres can be combined with cement, heavy pre-treatment is required to break up the blocks of fibres into thin fibre units of an effective diameter of 1 μm or less. Edge runners or hammer mills are used and because of their crystalline structure, the fibres tend to split into ever thinner fibres or cohesive bundles of parallel fibres without significant shortening of the fibre length. The fibres can then be easily dispersed in a water and cement suspension by continuous dilution and mechanical stirring. It is found that asbestos has a curious affinity for Portland cement which settles on the surface of the fibres and tends to remain there even under high dilution or water extraction. This enables the water content to be reduced from about 90 per cent to 20 per cent without segregation of the cement and leads to a very well distributed and bonded fibre composite.[8] This

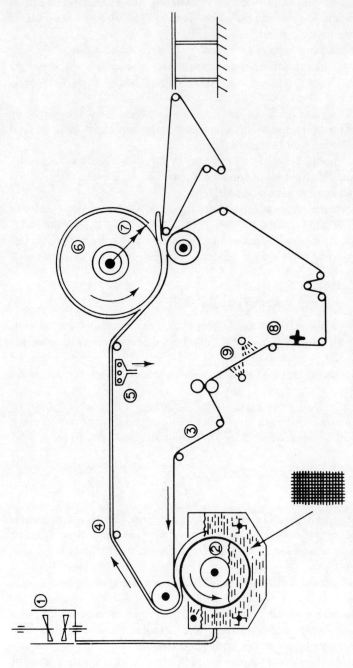

Figure 9.2. The fabrication of asbestos-cement sheets by the Hatschek process.

1. Mixer, agitator
2. Screen cylinder
3. Felt band
4. Ply of asbestos cement
5. Dewatering
6. Calender
7. Cutter
8. Beater
9. Sprayer

(Reproduced from *Technical Report No. 1 (51.067)*, July 1973, pp. 18–19, by permission of The Concrete Society) property forms the basis of the most widely used method of manufacture known as the wet process or Hatschek process.

9.2.1.1 Hatschek process[9]

The essential steps of the Hatschek process, which was developed from paper making principles in about 1900, are schematically shown in Figure 9.2.

A slurry, or suspension of asbestos fibre and cement in water at about 6 per cent by weight of solids is continuously agitated and allowed to filter out on a fine screen cylinder. The individual plies on the cylinder are about 0.6 mm to 1.4 mm thick and are transferred by a moving felt band to a calender where they are built up to the required thickness before being cut with a knife and transferred to a platform for subsequent processing.

The dispersion of the fibres is virtually in two dimensions not only because of the plied structure but also because of the rotation of the primary filter which gives a predominant alignment of the fibres in the direction of rotation, a feature which is commented upon in Section 9.3.2 on the physical properties of the composite.

In its fresh state, asbestos-cement has a considerable tear strength and can be readily handled, shaped on formers and cut to length using a knife.

9.2.1.2 The Mazza process[6]

The Mazza process is used to make asbestos cement pressure pipes and is a modification of the Hatschek process. The dewatered plies are transferred to a rotating steel mandrel instead of the calender and successive layers are compacted by pressure. The fibres are preferentially oriented in the circumferential direction in order to increase the bursting pressure of the pipes which are cured in water after removal of the mandrel.

9.2.1.3 The semi-dry or Magnani process

In the Magnani process, the solids to water ratio is in the neighbourhood of 0.5 and the mix is heated and pumped onto a belt where it is spread and levelled by reciprocating rollers. Both the belt and the rollers may be shaped to form corrugated or profiled sheet and vacuum boxes under the belt move with it to suck excess water from the hot mix.

9.2.1.4 The Manville extrusion process[6]

The asbestos fibres, cement, fine silica, and a plasticizer such as polyethylene oxide are fed from a hopper into a mixer with just sufficient water to produce a stiff mix. The mix is forced through a steel die by a worm drive to give extruded sections of the desired profile.

9.3 PROPERTIES OF ASBESTOS CEMENT

9.3.1 Standard Specifications

Asbestos cement is the only fibre composite discussed in this text for which there are International and British Standards requirements for certain properties. The German specification requirements have been listed by Klos[2] and some of the relevant British Standards are noted in References 10 to 12. The British Standard requirements for material properties are defined in terms of minimum bending strength, density, impermeability, and frost resistance and in addition, there are standard requirements for the performance of components such as corrugated or flat sheets, cable conduits, troughs, pipes, fittings, and cisterns.

The minimum bending strength varies between 15.7 MN/m^2 and 22.5 MN/m^2 depending on the type of sheet and whether it is semi- or fully-compressed. Also, the imposed loading to be carried on sloping or flat roofs varies between 0.75 kN/m^2 and 1.5 kN/m^2 depending on the roof slope and access conditions.

Water absorption should not exceed 20 per cent to 30 per cent of the dry weight depending on the type of product.

9.3.2 Tensile properties

The tensile properties of seven specially fabricated asbestos cements have been studied by Allen.[13] The manufacturing process imparts a certain degree of orientation to the fibres (Section 9.2.1.1) and therefore the properties were measured parallel to and at right angles to the direction of preferential fibre alignment. The fibre volume fraction and the length of the fibres were varied in the seven types and the results are shown in Table 9.1.

Types 2 to 5 in Table 9.1 contained similar fibres at an increasing fibre content. The void content increased with fibre content with a resulting decrease in modulus and increase in strain to failure. At high void contents the fibre−matrix bond strength is probably reduced which may account for the relatively low tensile strength of Type 5. Values for bond strength between 0.88−3 MN/m^2 have been quoted.[1] Types 1, 3, and 7 had short, intermediate and long fibres respectively but the effect of fibre length on the material parameters is not clear due to the interrelated effects due to change in void content.

Types 8 and 9 shown in Table 9.1 are commercial grades of asbestos cement and the strong directional effect is again shown in these materials. The moduli and tensile strengths of the fully compressed material are considerably higher than the semi-compressed sheet presumably due to a reduction in the void content. Tensile stress−strain curves for various asbestos cements have also been measured by Allen[14] and three types of curve are shown in Figure 9.3.

Curves A and C are for semi-compressed and fully compressed sheet respectively loaded parallel to the direction of preferential fibre alignment whereas curve B is for fully compressed sheet tested at right angles to that of curve C. All the curves show the characteristic feature of smoothness and there was no audible or visual

Table 9.1. Properties of seven asbestos cements (Data from Allen, *Composites*, June 1971, pp. 98–103, by permission of I.P.C. Science and Technology Press Ltd., Also, *Report 55*, September 1975, by permission of CIRIA)

Type	Fibre volume (per cent)	Void content of matrix (per cent)	Youngs Modulus GN/m^2		Ultimate tensile strength MN/m^2		Strain at failure × 10^{-6}	
			L^a	T^b	L	T	L	T
1	5.70	17.9	16.93	16.6	17.8	15.0	1280	1190
2	2.91	14.3	17.29	17.55	14.6	10.8	950	700
3	5.10	26.7	16.09	13.54	20.3	12.3	2320	1540
4	7.32	32.3	14.73	13.82	25.4	18.4	3700	2280
5	14.85	60.4	8.45	8.81	21.3	18.5	5060	4000
6	6.02	12.8	20.41	18.77	14.5	11.3	860	790
7	4.72	32.6	13.05	–	16.1	–	2540	–
8	Commercial Semi-compressed		13.6	15.2	16.1	9.5	2110	700
9	Commercial Fully compressed		25.6	25.0	27.1	17.2	2130	770

$^a L$ Load applied parallel to direction of preferential fibre alignment.
$^b T$ Load applied at right angles to direction of preferential fibre alignment.

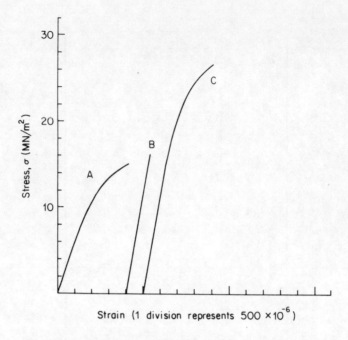

Figure 9.3. Tensile stress–strain curves for asbestos-cement (Reproduced from Allen, *Report 55*, September 1975, by permission of CIRIA)

cracking even when examined under load by low-powered microscopy. This type of behaviour is described theoretically in Section 3.4.2 and is quite unlike that of glass-reinforced cement. However, Allen suggests that the stress–strain curves are consistent with the progressive development of very fine cracks at strains in excess of the normal ultimate strain of the matrix even if there is no direct proof that this is so.

9.3.3 Properties of commercial asbestos cement

A range of properties for fully compressed and semi-compressed asbestos cement sheets as quoted in the manufacturers literature is shown in Table 9.2. The effect on the tensile and flexural strengths of the direction of stress relative to the fibre direction is clearly shown in the table. The compressive strength and modulus of elasticity however are not greatly affected by fibre orientation and these properties are likely to be more dependent on the density, although published data is scarce.

Few strength properties are available for semi-compressed sheets, possibly because the effect of manufacturing technique and density can lead to a wide range of properties but the moduli of rupture are generally quoted as being a minimum of 16 MN/m^2.

Table 9.2. Properties of asbestos cement. Data taken from manufacturers' literature

		Fully compressed flat sheet Autoclaved or steam cured	Semi-compressed Flat sheet
Density Kg/m^3		1,800 to 2,100	1,200 to 1,400
Tensile strength MN/m^2	Parallel to fibre	19 to 25	Actual
	Right angles to fibre	12 to 17	
Modulus of rupture MN/m^2	Parallel to fibre	43 to 59	Properties
	Right angles to fibre	30 to 43	
Compressive strength MN/m^2	Parallel to fibre	50 to 190	rarely
	Right angles to fibre		
Modulus of elasticity GN/m^2	Parallel to fibre	24 to 25	quoted
	Right angles to fibre		
Thermal Expansion x 10^{-6} per °C		9 to 10	10
Water Absorption Percentage of dryweight		14 to 18	30 to 40

9.3.4 Impact resistance

The impact resistance of asbestos cement is notoriously low and this may be due, in part to the stiffness of the material in the post matrix cracking zone such that imposed deflections caused by shock loads will cause very high local stresses and hence fracture. Contributory factors are short fibres with little energy being absorbed by fibre pull out.

Fracture energies under impact of conditions of 2 N mm/mm^2 have been quoted[6] which is close to that of the matrix alone.

9.3.5 Durability

Asbestos cement is known to be very durable under natural weathering conditions and Jones[15] has reported that no deterioration in flexural properties takes place due to weathering but that the material becomes progressively more brittle.

However, Opoczky and Pentek[16] have examined asbestos cement sheets at ages of 2, 16, and 58 years and have shown that asbestos fibres do suffer a certain amount of corrosion which is compensated for, in terms of composite strength by an increase in bond between the fibre and the cement. The corrosion of the fibre is promoted by the presence of airborne carbon dioxide which causes surface carbonation of the fibre. Also, certain magnesium hydroxides and magnesium carbonates may be formed as reaction products.

Attempts have been made by Majumdar et al.[3] to measure the strength of fibre bundles extracted from the asbestos cement boards stored under various exposure

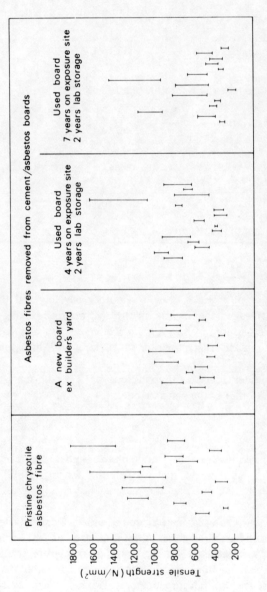

Figure 9.4. Tensile strength of asbestos fibres. Fibre bundles twisted to form a taut rope before measurement of diameter. Bars represent values based on minimum to average values (Reproduced from Majumdar, West, and Larner, *Building Research Establishment Current Paper CP 24/77*, May 1977, by permission of The Controller, HMSO. Crown copyright reserved)

conditions. The measured strengths are shown in Figure 9.4 and it is apparent that the variability in strength, both in the pristine condition and when extracted from boards is sufficiently great to mask any small variations in strength which may exist.

9.4 CONCRETE REINFORCED WITH ASBESTOS FIBRES

Chrysotile asbestos fibres have been used in concrete at up to 3 per cent by weight of cement by Winer et al.[17] The workability was considerably reduced with little or no gain in strength properties and as a result the material did not appear promising for construction purposes.

REFERENCES

1. Hodgson, A. A., 'Fibrous silicates,' *Lecture Series No. 4*, Royal Institute of Chemistry, London, 1965.
2. Klos, H. G., 'Properties and testing of asbestos fibre cement', *Fibre-reinforced Cement and Concrete*, RILEM Symposium, Vol. 1, 1975, pp. 259–267.
3. Majumdar, A. J., West, J. M., and Larner, L. J., 'Properties of glass fibres in cement environment', *Building Research Establishment Current Paper*, CP 24/77, May 1977.
4. Department of Employment, *Asbestos: Health Precautions in Industry*, 'Health and Safety Executive,' Report 44, H.M.S.O. 1975.
5. Gilson, J. C., 'Health hazards of asbestos,' *Composites*, Vol. 3, No. 2, March 1972, pp. 57–59.
6. Ryder, J. F., 'Applications of fibre cement,' *Fibre-reinforced Cement and Concrete*, RILEM Symposium, Vol. 1, 1975, Construction Press Ltd., pp. 23–35.
7. Asbestos Products Association, *Manual of Testing Procedures for Chrysotile Asbestos Fibre*, 1962.
8. Krenchel, H., 'Fibre-reinforced brittle matrix materials,' *Fibre-reinforced Concrete*, A.C.I. Publication SP-44, 1974, Construction Press Ltd., pp. 45–77.
9. Concrete Society, 'Fibre-reinforced cement composites,' *Technical Report*, 51.067, July 1973, pp. 18–19.
10. British Standards 4624 (1970), *Methods of Test for Asbestos and Asbestos--Cement Building Products*.
11. British Standards 690, Part 3: 1973, Part 4: 1974, *Specifications for Asbestos–Cement Plates and Sheets*.
12. British Standards 5247, Part 14: 1975, *Code of Practice for Sheet Roof and Wall Coverings. Corrugated Asbestos Cement*.
13. Allen, H. G., 'Tensile properties of seven asbestos cements,' *Composites*, June 1971, pp. 98–103.
14. Allen, H. G., 'Glass-fibre reinforced cement: strength and stiffness,' *CIRIA Report 55*, September 1975.
15. Jones, F. E., 'Weathering tests on asbestos cement roofing materials,' *Building Research Technical Paper*, No. 29, London H.M.S.O. (1947).
16. Opoczky, L., and Pentek, L., 'Investigation of the "corrosion" of asbestos fibres in asbestos cement sheets weathered for long times,' *Fibre-reinforced Cement and Concrete*, RILEM Symposium 1975, Vol. 1, pp. 269–276.
17. Winer, A., and Malhotra, V. M., 'Reinforcement of concrete by asbestos fibres,' *Fibre-reinforced Cement and Concrete*, RILEM Symposium 1975, Vol. 2, pp. 577–581.

Chapter 10
Fibres other than Asbestos, Glass, Polypropylene, and Steel

10.1 AKWARA FIBRES

In many parts of the world manufactured fibres may not be readily available and studies have therefore been made of natural vegetable fibres. Akwara (piassave fibre) is a natural vegetable stem fibre which is readily available in Nigeria and has been studied by Uzomaka[1] with regard to its use in concrete. The modulus of the fibre is low (1 to $4GN/m^2$) but it is stated to be dimensionally stable in water and durable in a cement matrix. However, at fibre volumes up to 5 per cent, there is no improvement in modulus of rupture of the concrete although the resistance to impact is increased.

10.2 ALUMINA FIBRES

The use of polycrystalline alumina filaments in a cement matrix has been studied by Bailey et al.[2] Filaments of 0.5 mm, 0.62 mm, and 0.77 mm diameter were used with strengths of 650 MN/m^2 and elastic modulus of 245 GN/m^2. Bond strengths at the initiation of pull-out varied between $4MN/m^2$ and $16MN/m^2$ and residual pull-out bond strengths of between $2MN/m^2$ and $10MN/m^2$ were measured. Tensile strengths of up to $4MN/m^2$ were achieved in the composite at fibre volumes of 3 per cent of continuous aligned fibres and these values were about three times the tensile strength of the unreinforced matrix.

10.3 CARBON FIBRES

Although carbon fibres are very expensive compared with most of the other fibres discussed herein, their qualities of high stiffness and tensile strength and a relative inertness to the alkaline conditions of cement paste have encouraged a number of workers[3-9] to attempt to establish the physical properties of carbon-fibre-reinforced cement.

10.3.1 Fibre production

Carbon fibres are available in various forms made from either continuous tows or short staples. Suitable organic materials in fibrous forms are carbonized at high temperatures and the resultant graphite crystallites are aligned by 'hot-stretching'. The strength and the stiffness of the fibre are dependent on the source material and the extent of hot-stretching, high modulus fibres being known as Type I and high strength fibres being known as Type II. By employing temperatures as high as 2750 °C during stretching, fibres with Young's moduli greater than 700 GN/m^2 have been produced. Textile fibres such as polyacrylonitrile or rayon, yield fibres of very good quality but those made from pitch and some types of agricultural waste have created a lot of interest on economic grounds.

Carbon fibres have a fibrillar structure similar to that of asbestos and their surface properties can be modified by suitable treatments. This is sometimes essential for controlling the interfacial properties of the composite. The fibres are inert to most chemicals but show anisotropy in physical properties such as thermal expansion.

10.3.2 Properties of carbon fibre cement

Ali et al.[3] have studied composites which were hand made by placing the reinforcement either in the tensile zone using continuous aligned fibres or in the body of the specimens using chopped strand mats random in a plane. Table 10.1 shows the properties of carbon fibre cement produced with high modulus fibres in comparison with matrix properties.

It can be seen from Table 10.1 that 3 per cent by volume of random fibre mat produces about a 70 per cent increase in tensile properties but the impact strength is decreased. However, a five fold increase in tensile strength can be achieved by 3.7 per cent by volume of continuous aligned fibres.

Tensile stress–strain curves for these materials are shown in Figure 10.1 from which it can be seen that the random mat composite has very little post-cracking ductility.

Table 10.1. Mechanical properties of high modulus carbon fibre cement composites (Reproduced from Ali, Majumdar, and Rayment, *Cement and Concrete Research*, 2 (2), 201–202 (1972) by permission of Pergamon Press Ltd.)

Fibre volume (per cent)	Fibre orientation	Young's modulus GN/m^2	Ultimate tensile properties		Impact strength kJ/m^2
			Stress MN/m^2	Strain $\times 10^{-6}$	
0	—	13.8	5.52	300–400	2
3.0	Random in plane chopped fibre mat	18.2	9.6	570	1.4–1.8
3.7	Continuous aligned	26.1	26.6	2160	3.6–4.5

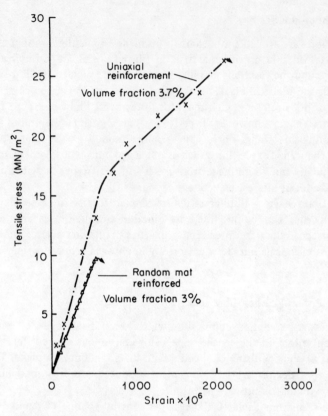

Figure 10.1. Stress—strain curves for carbon-fibre-reinforced cement composites in tension (Reproduced from Ali, Majumdar, and Rayment,[3] *Cement and Concrete Research,* **2**(2), 201—212 (1972) by permission of Pergamon Press Ltd.)

The durability of these composites was assessed by measuring the degree of strength retention by the samples which were kept under water at 18 °C and 50 °C over various periods of time. Very little change in strength was observed up to one year. This trend was confirmed by tests on composites produced by the spray-dewatering method[4] in which two different fibre lengths (11 mm and 32 mm) were used at 0.6 per cent to 1.3 per cent by weight. The samples were kept in different environments and under natural weathering conditions prevailing at the Building Research Station. Composites containing 1.3 per cent by weight of 11 mm long fibres gave, after one year a modulus of rupture value of 33 MN/m² and Izod impact strength of 4kJ/m². These values were slightly higher than the corresponding seven-day figures. Again, a continuous storage of the specimens under water at 60 °C did not have any significant effect on strength up to one year.

Aveston *et al.*[5] have studied the mechanical properties of carbon fibre cement in some detail using two types of composites. In aligned composites, which were made

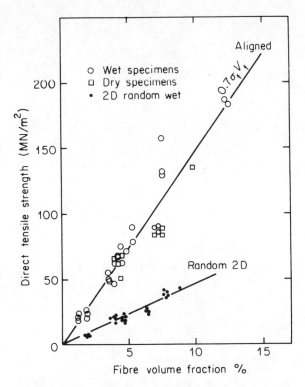

Figure 10.2. Ultimate tensile strength of carbon-fibre-reinforced cement (Reproduced from Aveston, Mercer, and Sillwood,[5] *Composites – Standards, Testing, and Design*, April 1974, pp. 93–103, by permission of the National Physical Laboratory. Crown copyright reserved)

by hand lay-up, a 100 mm wide thin veil (made from 10,000 filament tows) of carbon fibre was used. Pseudo-random composites were prepared from these sheets by rotating the mould through 10° for each layer of carbon. The tensile strength results of these composites are shown in Figure 10.2 and the corresponding stress strain relationships in Figure 10.3. In Figure 10.4 the relevant information on the initial modulus is given. It is clear from these results that enormous improvements in the strength and stiffness of cement can be secured by carbon fibre addition although similar improvements in stiffness can easily be achieved more cheaply by the inclusion of a high aggregate volume.

It has been suggested[10] that the failure strain of the matrix can be increased if a sufficient quantity of stiff fibres are uniformly dispersed in it and the type of stress–strain curves shown in Figure 10.3 could be taken to confirm this theory. However, as for asbestos cement composites, the onset of microcracking is very difficult to detect from changes in the slope of the stress–strain curve and more precise measuring techniques are required before the theory is fully confirmed.

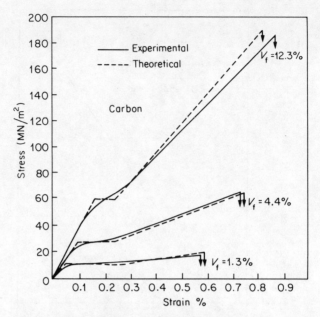

Figure 10.3. Tensile stress–strain curves for continuous carbon fibre reinforced cement (Reproduced from Aveston, Mercer, and Sillwood, *Composites – Standards, Testing, and Design*, April 1974, pp. 93–103, by permission of the National Physical Laboratory. Crown copyright reserved).

Sarkar and Bailey[8] have used continuous carbon fibres, Type Grafil A produced by Courtaulds, in volumes up to 10 per cent with a fine Portland cement of specific surface 800 m^2/gm and their conclusions confirm the work by Aveston.[5] Also Waller[9] has achieved moduli of rupture up to 250 MN/m^2 at fibre volumes up to 13 per cent. Briggs et al.[7] have produced additional information on several other engineering properties. Using aligned composites prepared by a filament winding technique they found that both bending and shear strength of carbon-fibre cement were considerably higher than those of the matrix but the compressive strengths decreased progressively as fibre contents increased. Both impact strength and work of fracture of the composites increased with increase in fibre content.

Briggs et al.[7] also reported very significant reductions in the drying shrinkage and creep of cement paste resulting from the addition of carbon fibres. A ten-fold reduction in shrinkage was measured in composites containing 5.6 per cent by volume of high modulus fibre stored in air of 60 per cent RH. In water storage the expansion of cement was similarly curtailed. Experimenting with composites having as low as 2 per cent by volume of fibre, the authors found that resistance to creep increased six fold when compared to that of cement paste. The resistance of carbon fibre cement to dynamic fatigue is indicated by the results shown in Figure 10.5. In these experiments bending loads were cycled at two frequencies, 30

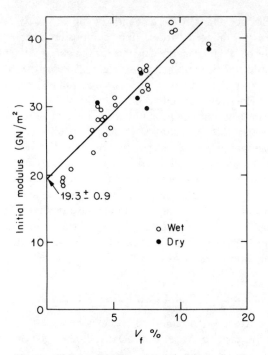

Figure 10.4. Initial Youngs Modulus of continuous carbon fibre-reinforced cement (Reproduced from Aveston, Mercer, and Sillwood,[5] *Composites − Standards, Testing, and Design*, April 1974, pp. 93−103, by permission of the National Physical Laboratory. Crown copyright reserved).

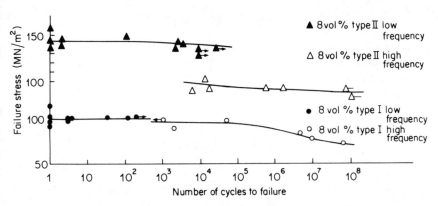

Figure 10.5. Fatigue life of carbon-fibre-cement containing 8 per cent by volume of fibres (Reproduced from Briggs, Bowen, and Kollek,[7] *Proc. 2nd International Carbon Fibre Conference*, 1974, by permission of Plastics and Rubber Institute)

and 2,000 cycles per minute. The tendency of the curves to level off around 10^8 cycles indicates that aligned carbon fibre composites containing 8 per cent by volume of fibre have fatigue limits in the region of 70–80 N/mm^2. These stress levels are much higher than the failure strength of the matrix and the results imply that matrix cracking at relatively low stresses may not be harmful to the composite performing at stresses below its fatigue limit.

Such optimism is not strengthened by the static fatigue results, also obtained by Briggs and coworkers. Composites were placed in different environments under static loads corresponding to 8–27 per cent of their ultimate strength for up to six months. Experiments were conducted in air and employing wet/dry and freeze/thaw cyclic variations. Considerable reductions in strength were observed in all conditions due to sustained loading even when the applied stress was below the limit of proportionality. However, the authors also observed that in the unloaded state the durability of the composites in all environments was excellent, those containing Type I fibres behaving particularly well. In this respect their conclusions are in agreement with those of Ali *et al.*[3]

10.4 CELLULOSE FIBRES

According to Krenchel,[11] cellulose fibres may be used in cement composites but volumes of 15 per cent to 20 per cent are required to give the composite suitable strength properties. Unfortunately, the fibres have the disadvantage that they are hygroscopic, the fibre dimensions vary with moisture content, the fibres rot if kept moist for long periods and they cannot tolerate heating to beyond about 100–120 °C.

It is therefore not likely that a direct replacement will be found for asbestos fibres by the use of cellulose but they may be included with asbestos fibres in asbestos cement production.

10.5 COCONUT FIBRES

Coconut fibres are very durable under natural weathering conditions and attempts have therefore been made to include them in cement based materials. Published data is sparse but they are likely to suffer from the usual disadvantages of vegetable fibres in that the modulus is very low and they are sensitive to moisture changes.

10.6 KEVLAR FIBRES

Kevlar is the proprietory name given to a group of high-modulus organic fibres developed by Dupont. PRD 49 and PRD 29 are examples, their properties being given in Table 2.1. The high strength and stiffness of the fibres imply a potential for the reinforcement of cement and Majumdar[12] described some work carried out at the Building Research Station on type PRD 49 in a preliminary study of material characteristics.

A modified spray-dewatering method was used to introduce a maximum of 2 per cent by volume of short fibres in random two-dimensional arrangements but the distribution of the fibres in the composite boards was far from satisfactory. Even with these incorporation problems the moduli of rupture of the composite were in the range of 40–50 MN/m^2 with ultimate tensile strengths between 14–16 MN/m^2. The resistance of these composites to dynamic fatigue was studied following the method developed by Hibbert and Grimer[13] for glass-fibre cement and the initial work indicated a fatigue life of 10^6 cycles at maximum applied stress levels of 20 MN/m^2. At ambient temperatures PRD 49 fibre cement composites were found to retain their strength properties over long periods of time but there is no firm data on this aspect to date.

Using a 70 per cent cement 30 per cent sand mixture as the matrix it was possible to autoclave Kevlar/cement composites at 180 °C for 16 hours. However, as Kevlar fibres are known to lose virtually all of their room temperature strength at about 300 °C and as their rate of creep also increases with temperature, the use of PRD 49 fibre cement composites in buildings will require a critical appraisal of each application. The behaviour of the material in fire together with a detailed study of physical properties has been described by Walton et al.[12a]

10.7 NYLON FIBRES

Nylon was one of the first of the polymer fibres to be included in cement and concrete but because of its relatively high cost compared with that of polypropylene, its commercial potential may be limited.

Goldfein[14] used up to 5.5 per cent by volume of nylon fibres in mortar and measured greatly increased impact strength and Williamson[15] confirmed the energy absorbing characteristics by explosive loading tests on slabs.

However, it was generally found[16] that the flexural strength of the composite was reduced by the inclusion of small volumes of short nylon monofilaments.

On the other hand, Walton and Majumdar[4] achieved moduli of rupture of up to 11 MN/m^2 using 4 per cent by weight (about 7% by volume) of 25 mm long nylon monofilaments and large increases in impact strength were observed which were not diminished by ageing. Both pre-mix and spray-up techniques were successfully used.

10.8 PERLON FIBRES

The effect of specific fibre surface on crack spacing in the composite has been described in Section 3.5 and Krenchel[17] has produced composites with a specific fibre surface of 25 to 30 mm^2/mm^3 using Perlon fibres with a diameter of 13 μm and a length of 6 mm. Perlon is a trade name which has been used for both nylons and polyurethanes, and the specific surface of the fibres used by Krenchel was sufficiently high that no discrete cracks were formed in the tensile zone in bending even at strains in excess of $20,000 \times 10^{-6}$. The fibre volume faction was about 8–9 per cent and the fibres were produced in the form of a paper through which cement paste was scraped.

Figure 10.6 shows the compressive and tensile strains on the surface of a beam in

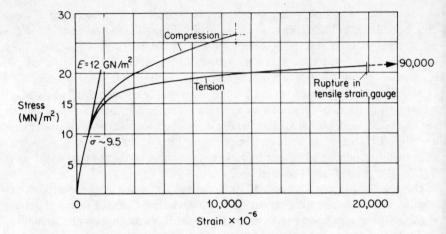

Figure 10.6. Bending test on a perlon fibre composite with a specific fibre surface of 25 to 30 mm^2/mm^3
Bending test with FRC-plate material:
Portland cement reinforced with perlon fibres
Reinforcement: 2–d random. l/d = 6.0/0.013, V_f = 8.6%
Matrix: w/c = 0.50,
(Reproduced from Krenchel and Hejgaard,[17] *Fibre-reinforced Cement and Concrete*, **2**, Discussion of Paper 7.4 (1975) by permission of the publishers, The Construction Press Ltd.)

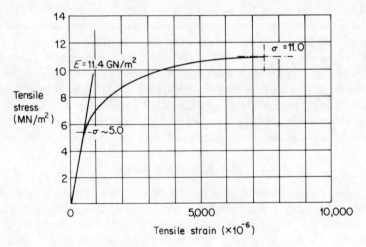

Figure 10.7. Tensile test on a perlon fibre composite with a specific fibre surface of 25 to 30 mm^2/mm^3
Tensile test with FRC-plate material:
Portland cement reinforced with perlon fibres
(Reproduced from Krenchel and Hejgaard,[17] *Fibre-reinforced Cement and Concrete,* **2**, Discussion of Paper 7.4 (1975) by permission of the publishers, The Construction Press Ltd.)

flexure and it can be seen that the neutral axis must move very near to the compressive surface at the ultimate load with the resulting high apparent modulus of rupture as described in Section 4.2.

The tensile stress—strain curve for the same material is shown in Figure 10.7 and the ultimate strength is seen to be about twice as high as the cracking strength of the matrix.

10.9 POLYETHYLENE

Polyethylene has not been used to any extent in cement composites because of its low modulus but the development of high modulus polyethylenes[18] with moduli of elasticity of up to 70 GN/m^2 has enabled the production of relatively cheap fibres with considerable potential in the field of cement based composites.

10.10 ROCK WOOL

Rock wool is manufactured from a molten rock mixture by blowing or spinning according to the same principles as used in the manufacture of glass wool. The fibres vary in length from 50 μm to 50 mm and they are very difficult to produce in a form suitable for impregnation with cement. Much of the basic work has been carried out by Krenchel[11,17,19] who has produced the fibres in the form of paper through which cement paste can be scraped to build up a laminated composite with properties equal to those of asbestos cement. However, the fibres are unlikely to be fully alkali resistant in cement paste and the long term properties of the composite are therefore suspect.

10.11 SISAL

A small study programme on the use of sisal fibres in concrete has been carried out at the Building Research Station.[20] The fibres tended to clump in the mix and setting was retarded by leaching or organic impurities from the presoaked cuttings but satisfactory pressed concrete beams were produced with up to 5 per cent of sisal by weight of cement. However, no additional strength was obtained by the addition of sisal reinforcement. More promising results have been obtained by Swift, at Kenyatta University College, Kenya, who has produced effective corrugated sheeting using combinations of long and short fibres.

REFERENCES

1. Uzomaka, O. J., 'Characteristics of akwara as a reinforcing fibre,' *Magazine of Concrete Research*, **28 (96)**, 162–167 (1976) September, *Discussion in Vol. 29*, No. 100, September 1977, pp. 161–164.
2. Bailey, J. E., Barker, H. A., and Urbanowicz, C., 'Alumina filament reinforced cement paste,' *Transactions and Journal of the British Ceramic Society*, **71 (7)**, 203–210 (1972) November.
3. Ali, M. A., Majumdar, A. J., and Rayment, D. L., 'Carbon-fibre reinforcement of cement,' *Cement and Concrete Research*, **2 (2)**, 201–212 (1972).

4. Walton, P. L., and Majumdar, A. J., 'Cement-based composites with mixtures of different types of fibres,' *Composites*, September 1975, pp. 209–216.
5. Aveston, J., Mercer, R. A., and Sillwood, J. M., 'Fibre-reinforced cements–scientific foundations for specifications: Composites–standards testing and design,' *National Physical Laboratory Conference Proceedings*, April 1974, pp. 93–103.
6. Waller, J. A., 'Carbon-fibre-cement composites,' *Civil Engineering and Public Works Review*, April 1972, pp. 357–361.
7. Briggs, A., Bowen, D. H., and Kollek, J., Paper 17, *Proc. 2nd International Carbon Fibre Conference*, The Plastics Institute, London, 1974.
8. Sarkar, S., and Bailey, M. B., 'Structural properties of carbon fibre reinforced cement,' *Fibre-reinforced Cement and Concrete*, RILEM Symposium, 1975, London, Vol. 1, pp. 361–371.
9. Waller, J. A., 'Carbon-fibre cement composites,' *Fibre-reinforced Concrete*, A.C.I. Publication, S.P. 44 1974, pp. 143–161.
10. Aveston, J., Cooper, G. A., and Kelly, A., 'Single and multiple fracture,' *Proceedings of Conference on the Properties of Fibre Composites*, N.P.L. 1971, pp. 15–26.
11. Krenchel, H., 'Fibre-reinforced brittle matrix materials,' *Fibre-reinforced Concrete*, A.C.I. Publication S.P. 44, 1974, pp. 45–78.
12. Majumdar, A. J., 'Discussion on properties of fibre–cement composites,' *Fibre-reinforced Cement and Concrete*, RILEM Symposium 1975, Vol. 2, pp. 605.
12a. Walton, P. L., and Majumdar, A. J., 'Properties of cement composites reinforced with kevlar fibres,' *Journal of Materials Science*, 13, May 1978, pp. 1075–1083.
13. Hibbert, A. P., and Grimer, F. J., 'Flexural fatigue of glass-fibre reinforced cement,' *Journal Materials Science*, 10, 2124–2133 (1975).
14. Goldfein, S., 'Plastic fibrous reinforcement for Portland cement,' *Technical Report 1757–TR*, U.S. Army Engineer Research and Development Laboratories, Fort Belvoir, Virginia, October 1963.
15. Williamson, G. R., 'Response of fibrous-reinforced concrete to explosive loading.' *Technical Report No. 2–48*, U.S. Army Corps of Engineers. Ohio River Division Laboratories. Cincinnatti, Ohio. Jan. 1966.
16. Monfore, G. E., 'A Review of fibre reinforcement of Portland cement paste, mortar, and concrete,' PCA Development Laboratories, September 1968.
17. Krenchel, H., and Hejgaard, O., 'Discussion of Paper 7.4: Can asbestos be completely replaced one day?' *Fibre-reinforced Cement and Concrete*, RILEM Symposium, 1975, Vol. 2, pp. 607–610.
18. Capaccio, G., and Ward, I. M., 'Preparation of ultra-high modulus linear polyethylenes; effect of molecular weight and molecular weight distribution on drawing behaviour and mechanical properties,' *Polymer* 15, April, 233–238 (1974).
19. Krenchel, H., *Fibre Reinforcement*, Academisk Forlag, Copenhagen, 1964.
20. Building Research Station, 'Poor outlook for sisal,' *Building Research Station News*, No. 14, Autumn 1970, pp. 25.

Chapter 11
Applications of Steel, Polypropylene, Glass, and Asbestos Fibres

No attempt will be made to give a complete list of applications because, in many cases, a number of projects of a similar type have been completed by different companies using similar materials. Only typical examples are therefore described giving, where possible, fibre types, quantities, and mix proportions.

11.1 STEEL FIBRES

The uses of steel fibre have mainly been in conjunction with concrete or mortar and little emphasis has been placed on the production of the thin sheet products which form the bulk of the asbestos cement or glass fibre cement industries.

A description of applications can be conveniently divided into those which involve site produced concrete and those concerned with pre-cast components. As is almost inevitable with the introduction of new materials many of the completed projects have been designed as trials and the satisfactory long term performance remains to be proven.

11.1.1 Site-produced concrete

11.1.1.1 Pavements

The main bulk outlets for steel fibre concrete have been in trials of highway and airfield pavement overlays or in full depth pavement applications. The basic principles behind the design procedure are outlined in Chapter 12 based on a paper by Williams[1] and the construction methods have been described in some detail by Lankard and Walker[2] and Lankard.[3]

Minor *in situ* work in the same field has included bridge-deck overlays in the U.S.A. and patching repairs to joints and spalled concrete surfaces in a variety of applications. The impact resistant properties have been utilized in expansion joint nosings for bridges which have been in service in the U.K. since 1972 when the first trial was carried out on the Upton Road Overbridge II on the M53 motorway.

Initially, a pre-bagged high alumina cement mix was used containing between 3 per cent and 4 per cent by weight of 0.38 mm diameter by 25 mm long round wire and a crushed rock fine aggregate, but in the mid 1970s the cement type was changed to Portland cement as a result of the restrictions on the use of high alumina cement. This product is known as Mono-joint produced by Industrial Flooring Services, Cheshire.

11.1.1.2 Industrial flooring

Warehouse and factory floors are applications which can utilize the increased impact resistance and post-cracking ductility of steel fibre concrete and several trials have been carried out to investigate the benefits which may result from the use of this material. An example is the concrete warehouse floor laid at Crossfields Ltd., Warrington, again by Industrial Flooring Services with pre-bagged mixes of similar proportions to the bridge nosings. A heavy duty floor was laid in 1971 for European Profiles Ltd., Llandybie Works, Ammanford, and in this case 6.5 per cent of weight of 0.38 x 35 mm long wire was used in a 2.4:1 mortar mix. Further examples include warehouse floors for Joylock Ltd., Cardiff, and for Quinton Hazell Ltd., Nuneaton using 3 per cent and 2 per cent by weight respectively of 0.38 mm diameter by 25 mm long Duoform* wire.

11.1.1.3 Hydraulic structures

Promising results have been obtained from the use of steel-fibre-concrete to resist cavitation, erosion, and impact damage in hydraulic structures such as sluice-ways and spillways associated with dams — Applications in the U.S.A. have been described by Hoff,[4] at the Libby Dam, Montana, the Lower Monumental Lock, and Dam on the Snake River, Washington and the Kinzua Dam, Pennsylvania. More than 1000 m^3 of fibre concrete was placed, some of it pumped, at fibre volumes varying from 1 to 2 per cent. Fibre dimensions included 0.25 x 0.56 x 25 mm long rectangular fibres and 0.41 mm diameter by 19 mm long round wires in mixes with maximum aggregate sizes of 20 mm and 10 mm respectively. Air entraining and water reducing admixtures were used in all cases. The batching, mixing, and pumping procedures for the Snake River job have been described by Kaden.[5] Although the results are not conclusive it appeared that the fibre concrete resisted erosion and cavitation very well in comparison with plain concrete and polymer materials.

11.1.1.4 Mining, Tunnelling, and rock slope stabilization

These applications have in common the stabilization of rock or loose surfaces by the use of steel fibre mortar or concrete applied by the gunite or shotcrete processes. In comparison with mesh reinforced gunite, one of the major benefits of

---

*Duoform is the subject of Patent No. 1235254 held by National Standard Co.

the fibre-reinforced material is said to be the considerable reduction in labour costs which results from the elimination of the necessity to pin the mesh to the rock surface. Also, a thinner layer of material can be used as it closely follows the rock profile rather than spanning cavities as with mesh techniques. However, considerable fibre wastage may occur due to rebound.

Trial applications were carried out by the National Coal Board in 1972 of gunited roadway linings at Bevercotes and Parkside collieries and British Rail have also carried out trials, for example the re-lining of a tunnel on the Leeds/Newcastle line using a 65 mm depth of fibre concrete containing 2 per cent by weight of dry material of 0.4 mm diameter by 25 mm long Duoform* wire.

In Sweden, the Besab technique (see Section 5.7) has been used to line shafts with a mortar mix containing 1.4 per cent by volume of 0.25 mm diameter, by 25 mm long fibres and a spraying application using this technique is shown in Figure 11.1.

The U.S. Army Corps of Engineers have successfully stabilized rock slope cuttings in Washington[3] using 1.5 per cent by volume of 0.25 mm diameter by 13 mm long fibres in a concrete shotcrete layer between 75 and 125 mm thick. Also, trials have been carried out at Newark, Lincolnshire in 1976 to gunite steel fibre concrete skins to form domestic houses.

11.1.1.5 Miscellaneous

A variety of one-off applications has included the 2-metre thick dome of the roof slab for the prototype fast reactor at Dounreay built by Taylor Woodrow Construction Ltd., which contained 2 per cent by volume of 0.5 mm diameter by 40 mm long wire to provide shear reinforcement in a confined area. Other projects have included heavy machinery foundations and crane rail support beams.

11.1.2 Precast components

Trials of many precast products have been made and it is probable that successful applications will rely on improved impact resistance and handling characteristics of the components or on the favourable flexural properties. The relatively small improvement in direct tensile strength over plain concrete means that the production of such items as concrete pressure pipes is likely to be inadvisable.

11.1.2.1 Manhole covers

Precast manhole covers and frames to replace cast iron components have been produced commercially and have received an Agrément Certificate. A 2:1 fine concrete mix with 7 per cent by weight of steel fibres was used in the prototype units and Duoform wire 0.40 mm diameter by 25 mm long is used in production units. A typical unit is shown in Figure 11.2.

*Trade mark of National Standard Co. Ltd.

Figure 11.1. Spraying steel fibre reinforced concrete using the BESAB technique (Reproduced by permission of the BESAB Company)

11.1.2.2 Slabs

One of the few structural applications has consisted of precast slabs about 1.1 m square and 65 mm thick supported by a tubular steel space frame to give a de-mountable car park at London Airport and this has performed satisfactorily from 1971.[7] Wire content was 3 per cent weight of 0.25 mm diameter by 25 mm

Figure 11.2. Steel fibre reinforced precast concrete manhole cover and frame (Reproduced by permission of the Cement and Concrete Association)

long fibres in a mix with 10 mm maximum size aggregate and the structure is shown during erection in Figure 11.3.

11.1.2.3 Marine applications

Also covered by the precast heading are marine applications such as weight coatings for undersea oil pipelines in the North Sea and Bredero Price have incorporated 0.35 mm diameter by 25 mm long Duoform fibre at their Immingham U.K. pipe coating plant. Other marine applications have included trials of 39 tonne dolosse breakwater units at the entrance to Humbolt Bay, California.[8] Fibres were

Figure 11.3. Steel fibre reinforced concrete slabs used in a car park at London Airport (Heathrow) (Reproduced by permission of Johnson and Nephew (Ambergate) Ltd.)

rectangular, 0.25 mm x 0.056 mm x 25 mm long and 10 dolosse were cast with 1.9 per cent by weight and 11 with 4.9 per cent by weight in mixes with up to 40 mm maximum size aggregate. Some problems were encountered with non-uniform fibre distribution and although the performance of the units was promising, it was concluded that fibre concrete was not yet a practical material for use in these units.

11.1.2.4 Pipes

Non-pressure pipes have been produced using most of the standard pipe making techniques although there is some doubt as to the economic advantages. For instance Henry[9] produced 1.52 m and 1.37 m diameter pipes but concluded that they could not meet the American Standard requirements when either an equal volume or an equal cost of fibres replaced conventional cage reinforcement. However, a Hungarian manufacturer[10] has found a commercial advantage in producing similar diameter pipes by the Packerhead process using about 2.3 per cent by weight of 0.5 mm diameter by 40 mm long Duoform wire. In this case the improvements were related to unreinforced concrete.

11.1.2.5 Refractories

A wide variety of commercial uses for refractory fibre concretes has resulted from the ability of the material to resist thermal and mechanical shocks at temperatures up to 1500 °C. Both stainless and carbon steel fibres have been used at volumes between 1 per cent and 2 per cent and the key to successful applications has been stated by Lankard[11] to lie in the conditions of temperature exposure. Good performance is said to be expected where the heating is one sided or where the refractory is subjected to rapid temperature excursions such as short plunges into molten steel. Similar effects have also been described by Nishioka[12] who used 10 per cent by weight of stainless steel fibre in a plate mill door 2.1 m wide by 2.8 m high subjected to temperatures of 1270 °C and impulsive loads for 4 months, the performance being stated to be much improved compared with conventional castables.

Applications in the U.K., using National Standard wire, in which the life of conventional components has been increased several fold included bogie car tops (600 °C) for Ross and Catherall Ltd., Killamash near Sheffield, charge hole covers (800 °C) for the British Steel Brookhouse Coking Plant, Yorkshire, ingot support blocks (1100 °C) for Jessop Saville Ltd., Sheffield, and stirring paddles (1500 °C) at British Steel Co. Stanton and Stavely Ironworks, Derbyshire.

11.1.2.6 Miscellaneous

Special items have included machine bases and frames and stairways. The impact and explosion resistant properties have been utilized in fillings for safes and security stores and for explosion resistant cladding panels for Gas Council compressor stations in 1974. These panels contained 5 per cent by weight of 0.4 mm diameter by 25 mm long Duoform wire in a mix with 1:2.25:2.25 cement:sand:10 mm aggregate.

Ferro-cement boat hulls are another area in which trials have been made but the corrosion resistant properties may well leave something to be desired in this application.

11.2 POLYPROPYLENE FIBRE

11.2.1 Chopped fibrillated film fibre

Fibrillated film fibre in the form of twine has been used in a variety of precast applications even though it is unlikely that the tensile or flexural strengths of concrete mixed in conventional machinery can be significantly improved by the addition of practical quantities (< one per cent by volume) of chopped fibre (see Chapter 3). The Shell designation 'Caricrete' is sometimes used for the fibre concrete.

The main advantages to the precaster are as follows:

(a) Improved impact resistance and the ability of the article to remain in one piece after being cracked;
(b) The low-fibre content of about 0.2 per cent by weight (0.5 per cent by volume) which has been shown to be effective;
(c) The low cost in comparison with steel or glass fibres;
(d) The reduction, or total absence, of steel reinforcement which allows savings in cost of steel and of labour for placing it;
(e) Savings in cost and weight caused by a reduction in thickness of products that had formerly needed cover for steel reinforcement;
(f) The resistance of polypropylene fibres to deterioration in a cement matrix or in aggressive environments;
(g) The simple nature of the alterations to the routines of mixing, placing and vibration, which are required in the factory or on site.

11.2.1.1 Pile shells

Wests Piling and Construction Co. Ltd., use a technique of driving a string of cylindrical shells threaded on a steel mandrel. The shells were formerly steel reinforced but have since 1969 been made of Caricrete after extensive impact testing under simulated field conditions but to a much higher degree of stress. The impact testing of these shells has been described by Fairweather[13] and in Chapter 7. The sections of 915 mm length and diameters between 280 mm and 533 mm, are manufactured in a highly automated plant and moulded under pressure and vibration at a high rate of production. More than half a million of these shells are made annually. The fibre concentration is typically 0.44 per cent by volume of 40 mm lengths of 12000 denier fibrillated twine.

Piles of this type being checked immediately after manufacture are shown in Figure 11.4.

Figure 11.4. Concrete shell piles containing chopped lengths of fibrillated polypropylene twine (Reproduced by permission of Wests Piling and Construction Co. Ltd.)

The successful use of polypropylene concrete in pile driving by West's has, of course, stimulated trials by other piling contractors, in particular those who make long precast piles with longitudinal prestressing wires. At the time of building the new Scunthorpe steel works, fourteen piles were driven, all made in the precasting beds on site, in which the helical steel binders were left out and replaced by random polypropylene staple fibres. Although it appeared that the manufacture and driving of the prestressed Caricrete piles presented no technical problems, there was no adequate economic incentive in this case to proceed with full production.

11.2.1.2 Cladding panels[14]

A series of panels was produced in 1967 for the Londonderry House Hotel on Park Lane, London, where in one area of the facade the normal specified thickness of 63 mm presented problems. The Caricrete panels without steel bars or mesh could be reduced to 33 mm thickness since the statutory cover to protect the steel reinforcement was no longer required. An extra bonus proved to be the ease of fixing the lighter weight polypropylene concrete panels.

The polypropylene concrete cladding for a building in Woking, Surrey, is shown in Figure 11.5. The large panels are 2.5 m by 1.5 m by 40 mm thick with aggregate

Figure 11.5. Concrete cladding panels containing chopped lengths of fibrillated polypropylene twine (Reproduced by permission of Cooper and Withycombe, Consulting Engineers, Guildford)

exposed on the ribs. The beams, 1.2 by 3.0 m are also 40 mm thick. Some steel reinforcement was included in top and bottom only to enable the green panels to be lifted out of the moulds after twenty-four hours. Here they were thickened to 75 mm. The areas of 40 mm thickness at the root of the ribs would have been 75 mm thick if conventional steel bars or mesh had been used.

11.2.1.3 Flotation units for Marinas[14]

Walcon Ltd. of Twyford near Winchester developed buoyancy units to carry jetties and walkways in harbours and marinas, with the emphasis on making them unsinkable, maintenance free, and comparable in cost to the steel floats or other types previously in use. Their trials resulted in the production of units, consisting of a core or expanded polystyrene, density approximately 16 g/1, surrounded by a casing of 18 mm thick polypropylene concrete, a typical installation being shown in Figure 11.6.

Manufacture comprises placing the block of expanded polystyrene with the help of spacers accurately in the centre of the mould, and pouring the concrete mixed with polypropylene fibres in the space between mould and shutter while external vibrators take care of the compaction. Various standard sizes are made, mostly 0.9 m deep, and with top surfaces between 1 m and 2 m by 1.5 m.

The increased impact strength which was discussed in Chapter 7 is one of the main features together with the impossibility of corrosive attack even after damage

Figure 11.6. Flotation units for Brighton Marina consisting of mortar containing chopped lengths of polypropylene twine surrounding an expanded polystyrene core (Reproduced by permission of Walcon Marine Ltd.)

has occurred. Since production started in 1971 Walcon pontoons have captured a major share of the market in the U.K.

11.2.1.4 Gunited polypropylene mortar[14]

Figure 11.7 shows an example of guniting the repair of a river wall in the tidal area of the Thames in 1968. Short staple of polypropylene twine, 20 mm long was mixed with a dry mortar and blown by air using standard gunite techniques to build up a total thickness of 90 mm at low tide. The advantages were that the fibre held the vertical mass of mortar together where steel mesh or chicken wire would otherwise have been needed. Also, the risks of subsequent rusting and spalling of the wall in this heavily corrosive area of the river bank were thus avoided and the repair was in perfect condition 8 years later.

Figure 11.7. Polypropylene fibres being used in a gunited repair of a river wall of the Thames 1968 (A Shell Photograph)

11.2.2 Chopped mono-filament

11.2.2.1 Decorative cladding

John Laing Research and Development Ltd. developed an aerated concrete composite with only 0.1 per cent to 0.2 per cent by volume of chopped polypropylene monofilaments added to a mix, designed with a view to creating a decorative sculptured finish, mainly for cladding panels. The new material was called 'Faircrete', which is a shortened version of 'fibre-air-concrete'.

Figure 11.8. 'Faircrete' cladding panels used on County Hall extension, Northallerton (Reproduced by permission of John Laing, Research and Development Ltd. and the County Architect, Northallerton)

Variation of the amount of air entrainment or of the type of aggregate used allows a choice of ultimate densities from 700 up to about 2,000 kg/m^3. Lightweight aggregates, ordinary gravels, or crushed rock aggregates may be used.

The material is described in Section 7.4.3 and a good example of the type of texture which can be achieved is shown in Figure 11.8.

11.2.2.2 Timber substitute

Another use for a modified form of Faircrete has been described by Pomeroy[16] in an attempt to find a substitute for timber crib supports for mining applications. Blocks were cast with 30 per cent porosity and 5 per cent by volume of 40 mm long polypropylene mono-filaments. Hoop reinforcement was provided by ductile steel and a good simulation of the rising load deflection curve of a typical timber construction was achieved.

11.2.2.3 Manholes[14]

Manholes up to 1.8 m in diameter and 7 m deep have been produced by Johnston Pipes Ltd., leaving out all steel reinforcement. Monofilaments were used in preference to fibrillated fibres because clients objected to the outside appearance of the fibrillated fibres whereas the monofilaments remained well embedded in the matrix. These products were used at the Gravelly Hill motorway interchange, Birmingham, England.

11.2.2.4 Sheet products

Trials have been made of the inclusion of chopped monofilaments in asbestos cement sheet products to improve the impact resistance and also, together with glass fibres, in the spray suction process to produce boards, with possibly improved long term ductility.

11.2.3 Continuous polypropylene ropes

Although it does not strictly fall into the category of fibre reinforcement, prestressed polypropylene rope is used in the production of concrete coated steel pipes by the German Company König Spezial–Tiefban. Ropes of between 4 mm and 8 mm diameter are used both longitudinally stretched and spirally wound in order to prevent permanent cracking due to undue deformation of the concrete coated steel pipeline.

11.2.4 Networks of continuous opened polypropylene film (Figure 7.2)

The main application areas are in cement mortars in thin sheet products where layers of continuous opened films are placed on top of each other, bonded by the matrix.[18]

In this form, very high impact strength can be achieved and pressed and moulded products may prove to be economical as direct replacements for asbestos cement sheeting. Possible applications include corrugated sheet roofing, cladding, troughs, rainwater goods, tunnel linings, pipes, crash barriers, and ventilation shafts.

It is also possible that these films may be used in boards with gypsum plaster as a matrix.

11.3 GLASS FIBRES

Although ropes of glass rovings coated with plastics have been used in place of steel reinforcement[19] and in place of prestressing cables[20] their use is outside the scope of this book which is mainly concerned with the utilization of closely-spaced fibres in more than one orientation.

The two main types of glass fibres which have been used in practical applications are borosilicate glass, also known as E-glass, in conjunction with low-alkali and high alumina cements, and alkali resistant glass, often known as Cem-FIL* in conjuction with ordinary Portland Cement.

However, in view of the unknown factors associated with the long term performance of both of these types of material, their use is likely to be limited to non-structural or semi-structural situations except in the case where the loading is very short term as in permanent formwork.

11.3.1 E-glass

Work on this material was initiated in the U.S.S.R.[21] in conjunction with low-alkali cements, and some prototype thin shell roofs of 6 m span and 5 mm to 7 mm thickness were produced in Kiev in 1963. Also, suspended ceiling panels were manufactured in high alumina cement with 10 per cent of glass by weight, the panel size being 6 m x 1.2 m x 10 mm to 15 mm thick. Other uses included internal waterproofing of cracked reinforced concrete tanks.

In the U.K., E-glass fibres were used in conjunction with high alumina cement from the late 1960s until about 1974 and products produced by Elkalite Ltd., included double skinned cladding panels with decorative finishes and insulating cores.[22] Also, pre-cast industrial chimney sections were marketed and British Rail installed some Elkalite heaters for points in the Southern Region. Other products included a 5-metre prototype workboat for Emsworth Yacht Marina, Hampshire, and houseboat support pontoons.

However, the restrictions imposed after 1975 in the U.K. on the use of high alumina cement in structures, and also the results of durability trials carried out by Allen,[23] which show considerable reductions in the tensile strength and strain to failure of E-glass in high alumina cement exposed to natural weathering, will probably limit the use of this matrix in the future.

Regardless of the generally accepted view that E-glass fibres are rapidly degraded

---

*Registered trade mark of Pilkington Brothers Ltd.

by Ordinary Portland Cement, structures have been made in the U.S.A. for a number of years[24] utilizing these materials, apparently without undue signs of distress.

11.3.2 Alkali-resistant glass

This glass has been developed to maintain its strength for long periods in Ordinary Portland Cement. However, there is still some doubt regarding the time scale during which the strength of the composite will reach a minimum limiting value and there is evidence[25] to suggest that the strain capacity in tension, the impact resistance, and to a lesser extent, flexural strength reduces continuously during periods of at least 5 years under natural weathering in the British Isles (Chapter 8).

Some of the non-structural or semi-structural components which have been used or are being developed using alkali resistant glass in conjunction with cement paste containing fine fillers are described below. Many of these products have been developed in conjunction with Pilkington Brothers Ltd., or Owens–Corning Fiberglass Corporation in the U.S.A., and a large number of companies have been licensed to produce fibre-cement composites in collaboration with the glass producers.

Although the fibres have been used in concrete at volumes of less than 1.5 per cent, problems of fibre breakdown due to prolonged mixing can occur and therefore most benefit can be gained from their use in cement or mortar at volumes between 3 per cent and 6 per cent and the main advantages for the user are noted below:
(a) The ability to be produced in thin sheet form which can be bent and moulded in the fresh state;
(b) The advantage that complex shapes can be produced by the spray technique or from pre-mixed material by pressing, injection moulding, or extrusion;
(c) Initially high flexural and impact strengths which can be utilized in manufacture, transport, handling, and erection of components in non-load bearing applications;
(d) A high degree of fire resistance.

11.3.2.1 Cladding panels

Single and double skin cladding panels are typically produced from the spray up process and are sized from 1 m x 2 m to 2 m x 4 m with skin thickness between 6 mm and 10 mm containing about 5 per cent by weight of glass fibres. They have been used to clad a variety of office and factory buildings in England including the Fibreglass Ltd. factory at St. Helens, Merseyside, the police traffic headquarters at Liverpool, a car park and shopping precinct at Stourbridge, office developments at Kingston on Thames, housing near Aberdeen and Inverness, Scotland, the RSPCA

Figure 11.9. Glass-reinforced cement cladding panels used on the Credit Lyonnaise Bank, London (Reproduced by permission of Whinney Son and Austen Hall, Architects and Designers, and G. Hana Ltd., Photographers)

building near Sydney, Australia, and London Headquarters of the Credit Lyonnaise Bank, Queen Victoria Street, London, the last being shown in Figure 11.9.

A certain amount of experience is required when designing panels with these materials and Soare and Williams[26] have described some of the problems which have been encountered in practice, such as manufacturing techniques, fixing details, waterproofing seals, and thermal insulation.

Figure 11.10. Bridge parapet units produced from glass reinforced cement by Thyssens (GB) Ltd. (Reproduced by permission of R. M. Douglas Construction Ltd., and E. W. Jinks, B.Sc., F.I.C.E., County Engineer and Surveyor for Mid Glamorgan County Council)

11.3.2.2 Permanent formwork

Glass-reinforced cement performs well in formwork because use can be made of its initially high flexural and impact strengths. Applications have included large waffle floor pans (1.5 m square by 1.6 m high) at Trumans Brewery, London, bridge deck formwork over the River Arun, near Littlehampton, and hexagonal earth retaining units on the M62 motorway in Yorkshire.[27]

Figure 11.11. Glass-reinforced cement units used to line sewers (Reproduced by permission of Charcon Composites Ltd.)

An example of permanent formwork produced by Thyssen (GB) Ltd. in use on a section of the M4 motorway (U.K.) is shown in Figure 11.10. The units enabled parapet sections to be cast without support from underneath and without subsequent making good.

Also in this category are lining segments for sewers and these have been used by the Greater London Council in a 1.75 m diameter sewer below the Portobello Road. High alumina cement was used in the production of the 10 mm thick segments with a sprayed glass-fibre content of about 5 per cent by volume compacted by hand roller. A variety of cement types is available and a typical installation by Charcon composites is shown in Figure 11.11. Units produced by the same company have also been ordered for the Bukit Timah contract in Singapore.

The use of glass reinforced sheet as combined shutter and surface reinforcement in composite concrete construction has been described by Dave.[28]

11.3.2.3 Small components

A variety of small components have been produced from pre-mixed material by extrusion or injection moulding to replace timber, cast iron, asbestos cement, or plastic sections. These include fence posts, pallets, window door frames, fire doors, ducting, drainage guttering, pipes, ceiling fascias, machinery covers,[27] noise barriers, road and garden furniture, and gas mains cover and frames.

11.3.2.4 Hydraulics and marine applications

Sheet piling has been produced by Charcon Ringvaart of Holland for use in canal revetments. Floating pontoons, consisting of a thin skin of glass reinforced cement applied over a lightweight core are also used in similar situations to those described in the section on polypropylene. Boat hulls, consisting of two 10 mm thick skins separated by a rigid foam polyurethane core have been undergoing trials for several years.[27]

11.3.2.5 Surface coatings

A technique of building blockwork walls by dry stacking the blocks and coating both the surfaces by trowelling on E-glass fibre reinforced ordinary Portland cement was developed in 1967 by the U.S. Department of Agriculture.[29] Further development work was carried out in the U.S.A.[30] using 4 per cent by weight of alkali resistant fibres and also in the U.K., where the material is available in the form of pre-bagged renderings containing about 3 per cent by weight of fibre. Minimum block thickness should be 75 mm and special attention should be paid to to stability of the blocks if higher than 3 m. Higher strength and more rapid construction than conventional blockwalls are claimed for the technique and a similar result can be obtained by spraying glass fibres and cement slurry.

In addition, a pre-bagged single coat rendering is available to weatherproof existing structures and an example of its use is on the Jack Kane sports and leisure complex at Craigmillar, Edinburgh, Scotland.

11.3.2.6 Miscellaneous items

These include a variety of items such as water tanks, swimming pools, low cost and refugee housing, and grain silos.

11.3.3 Glass Fibres in Concrete

Applications in concrete are rare, but concrete pavement overlays 50 mm and 75 mm thick were carried out in 1974 in St. Paul, Minnesota[30] using 1.36 per cent by volume of alkali resistant fibre. The mix proportions by weight were 1:1.63:1.38:0.67 cement:sand:10 mm aggregate:water. Another experimental project was to construct the floor slab of a residential house with glass-fibre concrete containing 1.5 per cent to 2 per cent by volume of 25 mm long fibres.[31]

Figure 11.12. Glass-reinforced concrete pipes (Reproduced by permission of ARC Concrete Limited)

Composite concrete pipes have also been produced commercially by ARC Concrete, near Bristol, England. The main proportion of the wall of these pipes consists of unreinforced concrete with glass fibre, in the form of continuous strands, being concentrated at the inner and outer surfaces. Among advantages claimed for the pipe are that it has a lower weight and an in-wall joint rather than the conventional bell end. (See Figure 11.12.)

11.3.4 Glass Fibres in Gypsum Plaster[32]

Glass-fibre-reinforced gypsum has been used in trials of a variety of components including school partitioning systems, flooring units, and fire barriers in cableways in Dungeness B Nuclear Power Station. Also glass-reinforced plaster moulds have been used in the casting of refractory special shapes where the life of some moulds has been increased from 30 to 60 casting.

11.4 ASBESTOS FIBRES

Asbestos cement products are so well-known in comparison with the other composites described herein that no attempt will be made to give a detailed description of all the applications. However, the text would be incomplete without a mention of the material which has, without doubt, reigned supreme for more than

Table 11.1. Deliveries of asbestos cement products in the UK in 1976 (Department of Environment, *Housing and Construction Statistics*, No. 20, 1976. Reproduced by permission of The Controller, HMSO, Crown Copyright reserved)

Product and application	Quantity (1000 tonnes)	Percentage of total
Corrugated sheets including roof decking	299	69
Flat sheets of all types	48	11
Rainwater and soil goods	7	2
All other products including pressure pipes	82	18
Total	436	100

half a century as the most successful example of a cheap, durable fibre, reinforcing a low tensile strength matrix.

In 1976 world asbestos production was about 5 million tonnes per year[33] and about 70 per cent to 75 per cent of this went into asbestos cement products. For instance, in 1976 in the UK alone, there were 436,000 tonnes of asbestos cement products delivered of which 69% was used for corrugated sheeting and accessories.[34] Due to the depressed state of the UK construction industry in 1976 this was a low figure compared with the early 1970s, the output in 1973 being nearly 40 per cent greater than in 1976.

Table 11.1 gives a breakdown of the most common components used in 1976. Roofing and cladding for agricultural and industrial buildings formed by far the largest application and the ability to be moulded into complex shapes has enabled a wide range of accessories, such as ridge pieces, cappings, eaves fillers, and flashings to be produced for the roofing applications.

Corrugated sheeting is also used in structural situations such as retaining walls for river banks and permanent shuttering for concrete. The sheeting may be coloured by surface coatings or pigments, and in the U.K. corrugation heights range from 15 mm to 95 mm with maximum purlin spacings of 1.67 m. However, in other countries, much greater corrugation heights and spans have been used.

Flat sheeting is a very versatile material with a range of densities from about 960 kg/m^3 for fire-resistant insulating board to about 1840 kg/m^3 for fully compressed sheet. A range of decorative finishes and moulded textures is available in compressed sheet but it is used more extensively for cladding houses and multi-storey buildings in the U.S.A. and Europe than in the U.K. Composite panels with two 6 mm skins of asbestos cement sheet separated by timber studding are used in the Isabest system for low cost dwelling in Israel.[35]

Flat sheet is also used for slates or diagonal tiles for light roof coverings with a weight of about 20 kg/m^2 and, with colouring pigments, they have been used to replace natural slate. Fully compressed sheet has a number of additional minor uses such as infil panels, shelving, bath panels, shuttering, dropping boards for poultry houses, bench tops, pig pen partitions, and weatherboard strips. Water based and chlorinated rubber based paints can be applied without a primer but gloss paints require an alkali-resistant primer.

Pressure pipes have been used for many years for conveying mains water, sewage, gas, sea water, slurries, and industrial liquors. Diameters commonly range from 50 mm to 900 mm with working pressures from 0.75 MN/m^2 to 1.25 MN/m^2 and, in addition, the crushing loads on buried pipes can be quite severe. An advantage of asbestos cement pipes is that their smooth uniform bore with freedom from formation of internal deposits means that the hydraulic resistance is low. Also, non-pressure fluid containers and pipes such as rainwater goods conduits, troughs, tanks, and flue pipes account for a large proportion of the minor applications of asbestos cement.

Extrusion processes are used in the U.S.A. and in Belgium to produce long lengths of hollow or open sections with complex profiles such as sills, coping, and decking units.[35]

The impact in the U.K. of the Asbestos regulations 1969 and the Health and Safety at Work, Etc. Act 1974 on the utilization of asbestos cement products is not yet clear. However, the Factory Inspectorate have recommended that basic precautions be taken during certain cutting operations, in particular the inhalation of dust from cutting, turning, or drilling asbestos cement should be avoided.[36] Preferably the cutting area should be dampened and waste should be buried or disposed of in such a way as to prevent dust being blown about. Clothing should be kept free from asbestos cement dust.

In addition, many asbestos cement products will be required to carry a Government Health Warning from October 1976.

REFERENCES

1. Williams, R. I. T. 'Steel-fibre concretes in pavements and pavement overlays,' *Short Course on fibre-cement and fibre concrete for practising Civil Engineers*, Paper No. 7, University of Surrey, England, April 1976, pp. 39.
2. Lankard, D. R., and Walker, A. J., 'Pavement and bridge deck overlays with steel fibrous concrete,' *Fibre-reinforced concrete*, ACI Publication SP-44, 1974, pp. 375–392.
3. Lankard, D. R., 'Fibre-concrete applications,' *Fibre-reinforced Cement and Concrete*, RILEM Symposium, 1975. Construction Press Ltd., London, pp. 3–19.
4. Hoff, G. C., 'The use of fibre-reinforced concrete in hydraulic structures and marine environments,' *Fibre-reinforced Cement and Concrete*, RILEM Symposium, 1975, Construction Press Ltd., London, pp. 395–407.
5. Kaden, R. A., 'Pumping fibrous concrete for spillway test,' *Fibre-reinforced Concrete*, ACI Publication SP-44. 1974, pp. 497–510.
6. Pomeroy, C. D., and Brown, J. H., 'Tailoring fibre concretes to special requirements,' *Fibre-reinforced Cement and Concrete*, RILEM Symposium, 1975, Construction Press Ltd., London, pp. 435–444.
7. Anon, 'Wire-reinforced precast concrete decking panels,' *Precast Concrete*, December 1971.

8. Barab, S., and Hanson, D., *Investigation of Fibre-reinforced Breakwater Armour Units*, ACI Publication SP-44, 1974, pp. 415–434.
9. Henry, R. L., 'An investigation of large-diameter concrete pipe,' *Fibre-reinforced Concrete*, ACI Publication SP-44, 1974, pp. 435–454.
10. Szabo, I., 'Applications of steel fibre-reinforced concrete,' *Fibre-reinforced Cement and Concrete*, RILEM Symposium, 1975, Discussion volume published 1976, Construction Press Ltd., London, pp. 483–486.
11. Lankard, D. R., and Sheets, H. D., 'Use of steel-wire fibres in refractory castables,' *Bulletin American Ceramic Society*, Vol. 50, No. 5, 1971, pp. 497–500.
12. Nishioka, K., Kakimi, N., Yamakawa, S., and Shirakawa, K. Effective application of steel fibre reinforced concrete. RILEM Symposium. Fibre-reinforced cement and concrete, 1975, pp. 425–433.
13. Fairweather, A. D., 'The use of polypropylene film fibre to increase impact resistance of concrete,' *Prospects for Fibre-reinforced Building Materials*, Building Research Establishment, 1972, pp. 41–44.
14. Zonsveld, J. J., 'Polypropylene fibre concrete,' *Course on Fibre-cement and Fibre-concrete for Practising Civil Engineers*, Lecture No. 4, University of Surrey, 1976, pp. 26.
15. Hobbs, C., 'Faircrete: An application of fibrous concrete,' *Prospects for Fibre-reinforced Construction Materials*, Building Research Establishment, 1972, pp. 59–68.
16. Pomeroy, C. D., and Brown, J. H., 'Tailoring fibre concretes to special requirements,' *Fibre-reinforced Cement and Concrete*, RILEM Symposium, 1975, Construction Press Ltd., London, pp. 435–444.
17. Walton, P. L., and Majumdar, A. J., *Cement-based Composites with Mixtures of Different Types of Fibres*, Building Research Establishment, CP 80/95, September 1975.
18. University of Surrey and Hannant, D. J., Provisional Patent Application No. 27371/76, *Improvements in or Relating to the Manufacture of Articles made from a Water Hardenable Mass and a Reinforcing Element*, July 1976.
19. Nawy, E. G., Newserth, G. E., and Phillips, J., 'Behaviour of fibre glass reinforced concrete beams,' *J.A.S.C.E. Structural Division*, 97, 2203–2215 (1971) September.
20. Soames, N. F., 'Resin-bonded glass-fibre tendons for prestressed concrete,' *Magazine of Concrete Research*, 15 (45), 151–158. (1963) November.
21. Biryukovich, K. L., Biryukovich, Y. U. L., and Biryukovich, D. L., 'Glass-fibre-reinforced cement,' *Budivelnik*, Kiev, 1964, Translation No. 12, C.E.R.A., London, 1966.
22. Addington-Smith, T. D., 'Glass-fibre-reinforced concrete – Applications,' *Concrete Society Symposium*, Birmingham University, September 1971.
23. Allen, H. G., 'Glass-fibre-reinforced cement – strength and stiffness,' *CIRIA Report No. 55*, September, 1975.
24. Ingerslev, E., 'General discussion,' *Fibre-reinforced Cement and Concrete*, RILEM Symposium, 1975, Volume 2. Discussion. Construction Press Ltd., London., London. Published 1976, pp. 632.
25. Building Research Establishment, *A Study of the Properties of Cem-FIL/OPC Composites*, CP.38/76, June 1976.
26. Soare, A. J. M., and Williams, J. R., 'The design of glass-fibre-reinforced cement-cladding panels,' *Fibre-reinforced Cement and Concrete*, RILEM Symposium 1975, Construction Press Ltd., London, pp. 445–452.
27. Ryder, J. F., 'Applications of fibre cement,' *Fibre-reinforced Cement and Concrete*, RILEM Symposium 1975, London, pp. 23–35.

28. Dave, N. J., O'Leary, D. C., and Saunders, J., 'The structural use of fibrous cement in composite concrete construction,' *Fibre-reinforced concrete*, ACI Publication SP-44, 1974, pp. 511–532.
29. Simons, J. W., and Haynes, B. C., 'Surface bonding – A technique for erecting concrete block walls without mortar joints,' *USDA Agricultural Research Service*, Athens, Georgia, No. CA-42-57, January 1970, pp. 6.
30. Pecuil, T. E., and Marsh, H. N., 'Fibre-glass surface bonding,' *Fibre-reinforced Concrete*, ACI Publication SP-44, 1974, pp. 363–374.
31. Anon, 'Experimental house uses fibrous concrete floor slab,' *Engineering News Record*, Vol. 193, 16th January 1975. 12 pp.
32. Ryder, J. F., 'Glass-fibre-reinforced gypsum plaster: Prospects for fibre-reinforced construction materials,' *Proceedings of an International Building Exhibition Conference*, London, November, 1971, Published by Building Research Establishment, 1972, pp. 69–89.
33. Turners Asbestos Fibre Ltd., *Personal communication*, August 1976.
34. Department of Environment, *Housing and Construction Statistics*, No. 20, 4th Quarter 1976, H.M.S.O., p. 64.
35. Ryder, J. F., 'Applications of fibre cement,' *Fibre-reinforced cement and concrete*, RILEM Symposium 1975, Construction Press Ltd., London, pp. 23–35.
36. Health and Safety Executive, Department of Employment, 'Asbestos: Health precautions in industry,' *Report 44*, H.M.S.O., 1975.

Chapter 12
Steel-Fibre Concrete in Road and Airfield Pavements

Contributed by:
R. I. T. WILLIAMS, M.Sc., C.Eng., MICE, F.Inst.H.E.
Reader in Concrete Materials Technology, University of Surrey

12.1 INTRODUCTION

This section on steel-fibre concrete in pavement applications has been separated from the main study of applications because the design of pavements and the solution of the many practical construction problems is a rather specialist topic which many engineers are not familiar with. The background to the subject and the review of design and construction of steel fibre concrete pavements have been taken from a paper by Williams.[1]

The structural design of conventional highway pavements in Britain[2] is based principally on the results of full-scale experimental sections[3] laid on heavily trafficked roads during the past 25 years. The procedure considers the cumulative number of commercial vehicles, defined as vehicles with an unladen weight of more than 1500 kg, to be carried by each slow lane during the design life. Tabulated factors, based on a statistical examination of data from weighbridges installed in a variety of roads, are provided for converting the cumulative number of commercial vehicles to the cumulative number of standard axles, a value of 8200 kg being used for the latter because this has the maximum damaging power when account is taken of both magnitude and frequency of axle loads. Separate design procedures are then followed for determining the nature and thickness of the materials to be used in concrete pavements and in blacktop pavements, the materials themselves and the construction procedures being specified elsewhere.[4]

For airfield pavements, less emphasis is placed upon frequency of application and the design method[5,6] concentrates instead upon taking account of the very heavy wheel loads and high tyre pressures of modern aircraft. The procedure involves calculating the equivalent single wheel load which has the same damaging power as the multiple wheels in the undercarriage assembly of a particular aircraft and then assigning a Load Classification Number (LCN) to the aircraft under consideration.

Design procedures based on experience, although adequate for conventional materials, may be unsatisfactory when new materials become available because the stress distribution and failure mechanism of the pavement may be different from that on which the empirical designs were based. A more fundamental analysis of stresses in layered systems such as that of Westergaard[7,8] or Croney[9] then becomes necessary in order that reduced pavement thickness or increased spacing of joints can be justified on a rational basis. Unfortunately, a standard theoretical procedure for such an ideal approach is not available and therefore engineers have resorted to judgment, or trial and error, in introducing steel-fibre concrete to pavement construction.

The properties of steel-fibre concrete which appear to be important from the point of view of pavement performance are increased flexural strength, improved post-cracking ductility, increased resistance to impact and repeated loading and improved spalling resistance and a number of small-scale field trials have been undertaken partly to gain construction experience with the new material and partly to obtain data regarding its behaviour under load. The considerable degree of success achieved in these trials has justified more comprehensive field trials and has already led to the material being used in a number of major schemes.

So far as new construction is concerned, the high cost of fibres has tended to discourage development and the position is unlikely to change unless a substantial reduction in slab thickness and an increase in joint spacing, relative to conventional concrete pavements, is shown to be justified by the experimental lengths now in service. Yet the prospect of laying concrete which contains its own reinforcement is very appealing so far as construction is concerned, so that the cost implications must be kept under continuous review.

The main interest is in the use of fibre reinforced concrete as an overlay material for restoring existing pavements. So far as concrete roads are concerned, the use of bituminous overlays has not always been entirely satisfactory and the escalation in the price of petroleum products has created a need to develop alternative materials. The finding that thin-fibre concrete overlays are performing well in trial sections is therefore welcome and it is interesting to note that overlays of this type have also been laid on existing flexible pavements. Developments have also taken place regarding the use of fibre concrete overlays in airports as a promising means of strengthening existing construction in order to meet the stringent requirements of the new generation of jumbo jet aircraft.

The present position is extremely encouraging but it must be emphasized that the developments have largely taken place during the past five years and that there are major gaps in knowledge which are at present dealt with principally by exercising engineering judgement. These points are referred to subsequently in the chapter.

12.2 MIX DETAILS

In comparison with conventional concrete mixes, the proportions used for fibre concretes differ in a number of respects which are dictated principally by the need to provide a higher proportion of mortar in the mix so that the required amount of fibre can be satisfactorily carried. The mixes are also laid in thin sections.

Consequently, the cement content is higher, the proportion of sand is higher and the maximum aggregate size is smaller. These factors are dealt with in more detail in Chapter 5.

The fibre content of the mixes is one of the variables at present being studied in test sections and, whilst there is economic merit in selecting a low value, many of the field trials have fibre contents in the range of one to two percent by volume. So far as the aspect ratio of the fibres is concerned, a high value is desirable from strength consideration whilst a low value reduces the risk of balling-up occurring during mixing and a satisfactory compromise is to use a ratio of 75:1 to 100:1. The diameter of the fibre is also a variable and, although satisfactory field performance has been reported with various sizes, recent work by Hannant[10] suggests that there is merit in using more substantial sizes in order to delay corrosion and rusting of wires spanning across cracks so that 0.5 mm diameter is likely to be a suitable size.

12.3 Construction Procedures

Detailed information regarding a considerable number of projects has been made available in the technical press during the past five years. Initially, difficulty was experienced in introducing the fibres into the mix but techniques have now been developed which overcome this problem. A tendency towards balling-up has been reported on occasion but, in general, this has been avoided by careful selection of fibre dimensions and by appropriate mix design.

A common observation is that mixing, placing, and finishing fibre concrete is very similar to plain concrete, using plant which is readily available. Projects have been successfully undertaken using semi-manual methods, concreting trains, or slip form pavers.

12.4 PAVEMENT DESIGN FOR NEW CONSTRUCTION

The most helpful guidance currently available is given by Rice[11] in connection with the use of fibre concrete in airfield construction. The design method involves analysing the proposed slab to ensure that the flexural stress induced is acceptable, determining the slab deflection since the use of a thin slab increases the risk of pumping the supporting layer, and ensuring that the stress developed in the supporting layer is low enough to preclude excessive permanent deformation. The full analysis involves the computation of stresses using the Westergaard formulae[7] and the elastic layer theory,[9] but a rule-of-thumb approach is included which is based on the fact that the inverse of the square of slab thickness is the dominant factor in the Westergaard analysis. Thus, approximately,

$$\frac{\text{Thickness of fibre concrete}}{\text{Thickness of plain concrete}} = \sqrt{\frac{\text{Working stress in plain concrete}}{\text{Working stress in fibre concrete}}}$$

On the assumption that the flexural strength of fibre concrete is twice that of plain concrete, and that the improved fatigue performance of fibre concrete allows a stress of 80 per cent of the strength to be applied in comparison with 75 per cent for plain concrete, the thickness ratio is approximately 0.7. But, to double the flexural strength, the volume fraction of fibres will probably need to be 1.5 to 2 per cent.

It is then recommended that one half of the required thickness of plain concrete should be used to begin the flexural stress computations for fibrous concrete. This is mainly for convenience but a second consideration is that there is less wheel stress interaction from an aircraft undercarriage assembly with a thin slab. The important point here is that the use of one half the thickness of plain concrete is for stress computation and not, as appears to be widely thought, for the final selection of the thickness of a fibre-concrete pavement. If no further analysis is contemplated, the use of a thickness ratio of 0.7 would appear to be more appropriate since this is supportable in terms of wheel load stresses.

For airfields, load transfer devices or thickened edge construction are recommended in order to avoid high edge stresses. Contraction joints to control shrinkage cracking are required and experience of fibre concrete pavements indicates that slab lengths of 30 m are acceptable[11] although this has not been generally confirmed.

Emphasis is placed on the need for a free draining course at least 100 mm thick under fibre concrete pavements because of the greater risk of pumping due to the increased deflection of thinner slabs.

12.5 PAVEMENT DESIGN FOR OVERLAYS

Again, reference is made to the paper by Rice[11] since there is no procedure available in the U.K.

The design is based on the established method used in America for determining the required thickness of conventional concrete overlay on an existing rigid pavement, and then modified to take account of the properties of fibre reinforced concrete. The calculation takes account of the condition of the base slab, the degree of bond between the slab and the overlay, and the thickness of the base slab.

Three interface bond conditions are envisaged:

(i) In the 'non bonded' condition, positive steps are taken to prevent bond from developing, usually by specifying a slip layer of polyethylene sheeting.
(ii) In the 'partially bonded' condition, no special precautions are taken either to ensure or to prevent bond, the preparation being minimal and often amounting to no more than sweeping the surface and possibly removing surplus sealing compound and loose material from the joints.
(iii) In the 'bonded' conditions, positive steps are taken to encourage bond by scabbling the surface and applying cement paste immediately before placing the overlay material.

12.5.1 Non-bonded

This is recommended for projects in which the base slab is in poor structural condition and where appreciable increases in structural capacity are required. It is also considered to be desirable when reflection cracking from the base slab is to be avoided but this is not mentioned in the design procedure.

The suggested formula is as follows:

$$h_0 = \sqrt{h_d^2 - Ch^2} \qquad (12.1)$$

where

h_0 = thickness of overlay

h = thickness of existing pavement

h_d = design thickness of plain rigid pavement

C = Condition factor = 1.0, existing pavement in good condition

= 0.75, existing pavement with initial cracks due to loading but no progressive cracks

= 0.35, existing pavement badly cracked or crushed

The thickness, h_0, thus determined is then reduced by one half for a fibrous overlay, the reduction not being due solely to the higher flexural strength of fibre concrete but also to its favourable post-cracking behaviour.

When a fibre-concrete overlay is to be placed on an existing flexible pavement, the recommendation is that is should be treated as a slab on-grade (the American term for new construction).

12.5.2 Partially bonded

This is recommended for applications where an increase in load carrying capacity is required and the base pavement is in fair condition.

The suggested formula is as follows:

$$h_0 = {}^{1.4}\sqrt{h_d^{1.4} - Ch^{1.4}} \qquad (12.2)$$

Where h_0, h_d, C and h are defined in equation (12.1) and the calculated value for h_0 is again reduced by one half for fibre concrete.

12.5.3 Bonded

Apart from the elaborate surface preparation required in this case, there is the further disadvantage that the fibre concrete is located in the compression zone of the slab so far as wheel loading stresses are concerned, and therefore its full potential cannot be realized. This is therefore envisaged only for situations in which surface defects are to be corrected with little or no increase in load carrying capacity. An alternative approach, not drawn from the design guide, would be to regard the function of the fibres in this case solely as a means of preventing local breakdown of the overlay in areas over which the required bond was not achieved, so that a low volume percentage of fibres might then be adequate.

Reverting to the design procedure, bond at the interface yields a thicker, monolithic slab. Even with full bond to produce a single slab, some shift of the neutral axis will occur because the overlay and the base slab have different properties.

The suggested formula is as follows:

$$h_0 = 0.9(h_d - h) \qquad (12.3)$$

where

h_0 = thickness of fibre concrete overlay (i.e. no further reduction is necessary)

h_d = design thickness of plain rigid pavement

h = thickness of existing pavement

12.5.4 Joints in overlays

When bonded overlays are used, the joints in the base pavement and overlay must be matched.

It is important to note that load transfer devices are required for non-bonded overlays.

12.6 EXAMPLES OF FIBRE CONCRETE CONSTRUCTION

The position has been comprehensively reviewed by Lankard[12] who gives details of significant projects completed between 1972 and 1975. This shows that there is some interest in new construction but that the main emphasis is on overlays for main roads, residential and city streets, airfield runways and taxiways, and parking areas.

The review is especially valuable since it emphasizes that fibre-reinforced concrete has emerged from laboratory status only since 1968 but makes it clear that considerable knowledge and experience has nevertheless already been gained. It is also reassuring since it draws attention to the fact that the projects include experimental lengths which will in due course provide a basis for a comparative study of a number of variables including concrete mix design, fibre type and quantity, joint spacing and design, overlay thickness, and type of bonding to the existing surface.

A particularly important comment in the paper is that the performance to date of steel fibre-reinforced concrete overlays, with a few exceptions at both extremes, has been satisfactory and that the user attitude is one of qualified satisfaction based on the as yet unknown long term performance.

On the 27th September 1968, a short length of fibre-reinforced concrete was constructed[13] as part of a road leading into the Machinery Division of the National Standard Company's plant at Niles, Michigan. This is probably the first pavement application of this new material and, since then, there have been numerous projects in a number of countries but notably in America.

In the following section of the chapter, brief details are given of some of the projects, these being chosen because of their scale, novelty of concept, unique performance, or particular relevance to practice in Britain.

12.6.1 Pavement performance investigation, Vicksburg

The U.S. Army Construction Engineering Research Laboratory carried out controlled traffic loading tests[14] on fibre reinforced concrete runway slabs in 1971 at the U.S. Army Waterways Experiment Station, Vicksburg, Miss. The loading simulated that of the C-5A which has a gross weight of 335 tons carried on three undercarriage assemblies, the two main units each having 12 wheels with four wheels on the nose gear. The simulator develops 13.4 tons on each of 12 wheels and was applied in sequence along five equally spaced parallel lines over a width of 5 m. Separately, parts of the test areas were subjected to simulated Boeing 747 traffic.

A plain concrete slab 15 m square and 254 mm thick, with joints dividing it into four bays, was constructed on a 100 mm sub-base of sand with a k-value of 35 kN/m^2/mm (CBR value about 4 per cent). This was subjected to 950 applications of the simulator and was reduced to a shattered condition with major structural cracks and severe spalling. It was then overlaid with 100 mm of steel fibre concrete, the interface condition being that of partial bond as it was prepared only by brushing and moistening. The simulator was again applied and the test discontinued after 6,900 loadings at which stage there was one working crack and hairline cracks.

Tests were also carried out on a 150 mm fibre reinforced slab 7.5 m x 15 m constructed on a 100 mm sand layer with a k-value of only 15 kN/m^2/mm (CBR value about one per cent). The transverse slab edges were thickened to 230 mm over a taper length of 0.75 m to reduce the free edge stresses to an acceptable level. This slab is effectively one-half the design thickness of the 254 mm slab when account is taken of the support conditions, yet it survived 8,735 loadings of the simulator.

In both the overlay and the on-grade fibrous slab, the only distress evident when testing was suspended was hairline cracking which would not interfere with aircraft operations on a pavement in use.

The fibre content was two percent by volume using 0.4 x 25 mm for the slab and 0.25 x 0.56 x 25 mm for the overlay. Manual techniques were used for laying, the material being delivered in ready-mixed trucks loaded to only 70 per cent of their capacity in order to improve mixing.

Instrumentation incorporated in the test showed that the fibre concrete slab on-grade transmitted a vertical stress of 70 to 80 kN/m^2 onto the sand layer and that the maximum deflection of the slab under a static application of the C-5A simulator was 5 mm. The strains developed were about 60 μs although much higher values of short duration were measured in the early stages of loading.

12.6.2 Tampa International Airport

Two fibre concrete overlay test sections were constructed[15] in February 1972 over 300 mm plain concrete slabs 7.5 m square which had been opened to traffic in

January 1966, but which began to show signs of distress in December 1966 and had subsequently suffered continuing deterioration under the increasing volume of jet aircraft.

One overlay was 150 mm thick, 23 m wide by 53 m long and was paved in three 7.6 m widths. The other overlay was 100 mm thick, 15 m wide by 15 m long and was paved in two 7.6 m widths. Construction was by CMI slipform paver with no provision made at the interface for either bonding or stress relief. The fibres used were 0.5 x 0.25 x 25 mm, the volume fraction being approximately 1.5 per cent. The handling characteristics of the fibrous concrete were found to be very similar to plain concrete. Some hairline cracks developed, mainly reflection cracks and particularly where the 100 mm overlay spanned the longitudinal joint in the base pavement. A subsequent report[16] stated that the cracks in the 150 mm overlay are non-working and hairline but that the cracks in the 100 mm overlay are 0.8 to 1.6 mm wide and show evidence of working.

12.6.3 Warren, Michigan

In October 1972, a fibre concrete overlay was placed[17] on a heavily trafficked urban interstate highway in Detroit. The existing reinforced concrete pavement, which was badly cracked, was simply swept and wetted and then overlayed using a CMI slipform paver. The overlay thickness was 25 to 76 mm, laid to a width of 15 m over a length of 335 m with transverse full depth joints being sawn after 24 hours curing at spacings of 15 m and 30 m and with longitudinal joints being cut in most of the 7.5 m widths. Fibre contents of approximately one and 1.5 per cent by volume were used, the fibres being 0.25 x 0.56 x 25.4 mm. Traffic was allowed on two lanes of the new pavement only 48 hours after construction.

Six months after construction, about two-thirds of the overlay was removed. The remaining section had some full width transverse cracks but was considered to be serviceable. Failure to achieve the design thickness of 76 mm and also the specified 2-day flexural strength of 4.5 MN/m^2 are understood to be the main reasons for the poor performance in this instance, but the suggestion has also been made that warm days following cold nights caused the only partially bonded overlay to curl and then to crack at the joints.

12.6.4 Greene County, Iowa

Lankard and Walker[17] regard this project as the most significant highway overlay experiment undertaken up to 1975. The construction took place in the autumn of 1973 on a three-mile section of highway E53 near Jefferson, Iowa.

It includes 33 sections of steel fibre concrete, each 122 x 6.7 m, laid to thicknesses of 50 and 76 mm. In addition, for comparison, there are four sections of continuously reinforced concrete pavement 76 and 100 mm thick and five sections of plain concrete and mesh reinforced concrete at thicknesses of 100 and 125 mm. The experiment also makes provision for mixes in which the binder is shrinkage compensated cement or cement/fly-ash.

Two types of fibre (0.25 x 0.56 x 25.4 mm and 0.64 x 63.5 mm) were used in amounts giving approximately 0.45, 0.75, and 1.2 per cent by volume. Two mixes were specified (356 and 445 kg/m^3 of cement), the maximum aggregate size being 10 mm and the sand content 50 per cent.

The original pavement was widened from 5.5 m to 6.7 m by laying a 0.6 m strip of lean concrete 100 mm thick along both edges. The surface condition covers bonded (by applying cement paste), partially bonded and non bonded (double thickness polyethylene sheet). The paving train consisted of a CMI spreader, a Quad City slipform paver and a Gomaco texturing machine.

Early observations suggest that the non-bonded sections are noticeably less prone to transverse cracks.

This is a comprehensive experiment but, unfortunately, it is only relatively lightly trafficked so that there will be a delay before meaningful conclusions can be drawn.

12.6.5 The M10 overlay

The trial was carried out[18] in May 1974 in order to evaluate the potential of steel fibre reinforced concrete as a thin overlay. The site chosen is on the northbound carriageway of the M10, a dual two-lane motorway, close to the roundabout at the beginning of the motorway. The daily flow of commercial vehicles in each direction is about 2,200, which is equivalent to some 1×10^6 standard axles per year.

The base concrete slab is 275 mm thick and 7.9 m wide with transverse dowelled expansion joints at 36.6 m intervals and reinforced with 5.4 kg/m^2 oblong mesh. Both expansion and longitudinal joints were formed by sawing, the latter being 3.7 m from the nearside edge. Delay in sawing the longitudinal joint sealing groove had led to a considerable amount of cracking roughly along the centre line of the existing carriageway. The concrete mix used had an aggregate/cement ratio of 7.9 and was not air-entrained and, in consequence, there was extensive surface scaling particularly at joints and cracks. The motorway was opened in November 1959 and a fairly considerable amount of surface repair work was subsequently undertaken to deal with the cracks and the scaling.

In the length to be overlaid, the average spacing of transverse cracks was about 1.2 m in the nearest lane and about 5.6 m in the offside lane; about one-third of the cracks had been repaired and, in many cases, cracks had developed in the repair patches. In addition, there was a more-or-less continuous longitudinal crack some 300 mm from the centre line of the carriageway. Nevertheless, after 15 years of heavy traffic, the overall structural condition was generally good.

In selecting the variables to be incorporated in the trial, an over-riding factor was the limit of 18 days during which the carriageway could be closed to traffic. In the event, overlay thicknesses of nominally 60 mm and 80 mm were chosen with 0.5 x 38 mm Duoform fibres at concentrations of 0.82, 0.67, and 0.4 per cent by volume, the base concrete being either just cleaned off or lightly scabbled, damped and coated with cement slurry. The existing transverse joints were matched in the overlay and, in addition, transverse contraction joints were formed at 12.2 m

Figure 12.1. Rear view of CPP 60 slipform paver laying steel fibre-reinforced concrete as an overlay trial on the M10 (Photo: R.I.T. Williams, University of Surrey)

spacing. The concrete mix had an aggregate/cement ratio of 4.3, using 10 mm maximum size aggregate and 50 per cent sand.

The material was laid through a CPP60 slipform paver, using wire guidance, in two equal widths so that the longitudinal joint in the overlay was not directly over the sawn joint in the base slab. The off-side lane was paved four days after the nearside lane with one track of the paver running in the centre reserve and the other on the nearside overlay which was protected by plywood sheets. Two views of the trial are shown in Figures 12.1 and 12.2.

The 28-day cube strengths of about 50 MN/m^2 and the 91-day flexural strengths of about 4.9 MN/m^2 were not significantly affected by the nominal fibre content and were rather less than had been expected on the basis of preliminary laboratory mixes. Fatigue tests, however, show that all the mixes tested had a considerably better performance than one would expect from plain concrete of equal strength.

Carefully undertaken surveys have traced the progressive development of cracks and, with a few exceptions, these have remained as hair cracks. After six months of heavy traffic, the best performance in terms of minimum surface cracking was from the thinner bonded sections. The authors emphasize the limited extent of the trial and, in concluding that the early performance is promising, suggest that the future behaviour may depend on how successfully the fibres spanning the cracks hold the individual pieces together and especially as water and salt penetrate into the cracks.

It is understood that cores subsequently extracted indicated good bond where the surface of the base concrete had been roughened but that the partial bond condition in fact amounted to the absence of effective bond.

Figure 12.2. The M10 trial showing a finished length of overlay on the slow lane and the base concrete, with a previously repaired crack, in the fast lane (Photo: R. I. T. Williams, University of Surrey)

12.6.6 Lorry Park, Wakefield

This project has been described in detail by McDonald,[19] briefly summarized by Kent[20] and referred to by Swamy and Lankard[21] in a comprehensive review of applications of steel fibre concrete. It is of particular interest, principally because it was undertaken in December 1970 and is therefore probably the first example of new construction on a fairly substantial scale.

The pavement forms a lorry park at the works of the Spencer Wire Division of Johnson and Nephew (Steel) Ltd., Wakefield. A total length of about 400 m was laid 2.7 m wide in lengths of 6 to 67 m at thicknesses of 12 to 100 mm, on a wet-lean concrete base. The fibre used was 0.25 x 25 mm, at concentrations of 0.9, 1.2, and 1.5 per cent by volume. Laying was through a CPP60 slipform paver. The exercise also included an area in which a 75 mm topping of fibre concrete was laid directly on a loose sand base.

The pavement has since been in continuous use and is reported[21] to have shown little cracking and to have given excellent performance and serviceability.

12.6.7 Fibre concrete pavement, Calgary

The project was undertaken in 1973 and details have been published by Johnston.[22] It is located on the University campus and forms part of a city bus route, with about 1,000 passes per week of buses weighing 9,280 kg unladen and up to 12,700 kg laden.

It is an extremely interesting project and makes provision in a 55 m length for a direct comparison of the performance of fibre concrete and plain concrete slabs. It is also new construction.

Fifteen slabs, each 3.65 x 3.35 m, have been laid in thicknesses ranging from 76 to 178 mm on a gravel sub-base with a CBR of 34.7 per cent. Brass coated steel fibres, 0.25 x 19 mm, were used in volume fractions of 0, 0.5, and 1.0 per cent.

The site itself is also of interest since the temperature varies from 32 °C in the summer to −40 °C in the winter, with frost penetration of 0.9 to 1.5 m and with only about 100 frost free nights per year. In addition, road grit and de-icing salts are employed to a considerable extent so that, overall, the environment is a severe one.

The principal conclusion which may be drawn from observations and detailed crack surveys during the first 12 months of use is that, for equivalent performance in terms of cracking, concrete reinforced with one percent of fibres by volume can be laid at 60 per cent and possibly as little as 50 per cent of the thickness required by plain concrete, the corresponding value being 75 to 80 per cent when the volume fraction is 0.5 per cent. It is also concluded that fibre concrete has performed significantly better than would be expected in terms of flexural strength alone, and Johnston[22] draws particular attention to the need for alternative or supplementary laboratory tests in order to evaluate the potential pavements.

12.6.8 West London Oil Terminal, Esso Petroleum Company

This is an outstanding example of the effective use of steel fibre reinforced concrete, although, as yet, no detailed information has been published regarding the project.

The existing paving is mainly of rolled asphalt on a cement stabilised gravel and, whilst the structural condition is satisfactory, a maintenance need had arisen due to the softening action of oil spillage from tankers.

A 100 mm steel fibre concrete overlay with 1.2 per cent by volume of 0.25 x 25 mm fibres has been laid on a polyethylene membrane over a very extensive area which is believed to be the largest steel fibre concrete pavement in the U.K., the work being started in October 1972. Although some bays developed transverse cracks or corner cracks, these do not appear to have deteriorated and the overlay is performing well. In addition, a new parking area has been constructed with a slab thickness of 150 mm and using the same type and concentration of fibre.

12.6.9 Fort Hood fibre concrete overlay

This was recently described by Williamson[23] as the first full-scale non-experimental steel fibre concrete placement and was completed in March 1974 at Fort Hood, Texas. It involved laying 2426 m^3 of fibre concrete as an overlay in a tactical equipment park used for the maintenance and repair of combat tanks and other tracked vehicles.

The existing pavement consisted of 120 to 180 mm of asphaltic concrete on a 180 mm lime stabilized base and had required replacement every three or four years due to the severe wear caused by the tracked vehicles, the tank for which the project was designed having a gross weight of 47,628 kg. The modulus of subgrade reaction was measured and its variability taken into account, and a computerized edge-loading Westergaard analysis undertaken to determine the flexural stress in the slab, the deflection of the slab and the subgrade stress under the load applied by the bogies of the tanks.

An overlay, reinforced with 1.5 per cent by volume of 0.3 x 12.7 mm fibres, was laid in widths of 6 m at thicknesses of 100 to 130 mm and transverse joints at 15 m spacing were sawn as soon as practicable to a depth of one-half the slab thickness. The operations of mixing, placing, and finishing were found to be very similar to those for plain concrete and fibre-balling was essentially non-existent due to the care taken in establishing a proper charging technique for the fibres. The usual means of texturing the finished surface by burlap drag was not suitable and a rollar system was used instead. Early evaluation indicates that the overlay is performing well with only one working crack in the entire project.

12.6.10 Applications associated with pavements

Lankard[12] gives details of the use of steel-fibre concrete since 1972 as an overlay material to existing bridge decks in thicknesses of 51 to 127 mm at fibre concentrations of 0.75 to 1.5 per cent by volume. Four of the projects described aimed at full bond with the existing deck by using epoxy and cement paste bonding agents, one was partially bonded and one was non-bonded. All the fully-bonded overlays and also the partially bonded overlay have developed transverse cracks which in most cases have remained tight whereas the non-bonded overlay has remained virtually crack free for three years. Details are also given of the use of fibre concrete in the State of Virginia for part of the construction of new bridge decks.

Steel fibre concrete has also been used to repair deteriorated portions of concrete slabs where, for example, spalling has occurred. Within this type of application is the use of 51 mm thick precast fibre concrete slabs for the rapid repair of unserviceable areas in tunnel decks and highway pavements.

A further application is in the construction of warehouse and factory floors, this being of relevance in this country in view of the interest at present being taken in adopting road construction practices for floor construction.

12.7 CONCLUDING REMARKS

There is little doubt that steel-fibre concrete has made a considerable impact on pavement technology and that, with a few exceptions, the early-life performance is very encouraging. However, it remains to be seen whether or not the performance over a longer period is satisfactory but this limitation is inherent when new materials or new concepts are introduced.

A study of the literature, together with subjective thoughts based on experience with other new materials, prompts the following observations:

(a) There is an urgent need for greater emphasis to be placed on using established analytical techniques for the structural design of pavements. Otherwise, continuation of a wholly empirical approach may lead either to unacceptable failures or to the non-use of potentially rewarding developments.

(b) For new construction and for overlays, research is needed to provide a more certain basis for selecting both the thickness of slabs and the spacing of joints.

(c) In a number of papers giving details of projects, it is reported that field strengths fall short of the values envisaged at the laboratory stage and suggests the need for greater emphasis either on mix design or on quality control, or both.

(d) There is growing evidence that strength alone does not identify the real nature and potential merit of steel-fibre concrete and research is needed to obtain a better understanding of its post-cracking properties.

(e) Many papers, and especially those relating to work in America, report the use of fly-ash as a partial replacement for cement. This has merit in terms of improved workability and therefore more reliable placing. There is also more widespread use of admixtures.

(f) Perhaps the most important need for development so far as materials are concerned is in relation to the cost of steel fibres. Recent advances in wire producing techniques may significantly reduce costs and, in addition, a greater willingness to select wire of more substantial dimensions, such as 0.5 x 50 mm or 0.6 x 60 mm, would reduce costs and at the same time provide greater insurance against deterioration by rusting and corrosion.

(g) A volume fraction of at least one per cent and preferably 1.5 per cent is probably desirable in the light of current experience and, in these amounts, the steel costs are very considerable.

REFERENCES

1. Williams, R. I. T. 'Steel-fibre concrete in new pavements and in pavement overlays.' *Fibre Cement and Fibre Concrete*, Course at University of Surrey, 1976, paper 7. 43 pp.
2. Road Research Laboratory, *A Guide to the Structural Design of Flexible and Rigid Pavements for New Roads*, 3rd Edition, Road Note 29, 36 pp. HMSO London, 1970.
3. Croney, D. and Loe, J. A., 'Full-scale pavement design experiment on A1 at Alconbury Hill, Huntingdonshire,' *Proc. Institution of Civil Engineers*, **30**, pp. 225–270 (1965) February.
4. Ministry of Transport, Scottish Development Department, *Specification for Road and Bridge Works*, Welsh Office, HMSO, London, 1969, pp. 195.
5. Directorate of Civil Engineering Development, *Design and Evaluation of Aircraft pavements 1971*, Department of the Environment London, pp. 19.
6. Directorate of Civil Engineering Development, *Aircraft Pavements: Standard Specification Clauses and Guide Notes. Series 500, Dry Lean Concrete: Series 800, Pavement Quality Concrete* Department of the Environment, London.
7. Westergaard, H. M., 'Stresses in concrete pavements computed by theoretical analysis,' *Publ. Rds.*, Washington, **7(2)**, 25–35 (1926).
8. Westergaard, H. M., 'Analysis of stresses in concrete roads caused by variations of temperature,' *Publ. Rds.*, Washington, **8(3)**, 54–60 (1927).
9. Croney, D., 'The design of concrete-road pavements: One-day meeting on concrete roads,' *Concrete Society Technical Paper PCS 20*, 1967, pp. 27.
10. Hannant, D. J., 'Contribution to discussion of paper 4.4,' *Fibre-reinforced Cement and Concrete*, RILEM Symposium 1975, Vol. 2, Discussion, The Construction Press Ltd., 533–538.
11. Rice, J. L. 'Pavement design considerations,' *Fibrous Concrete, Construction Material for the Seventies*, Conference Proceedings M-28, CERL, December, 1972, pp. 159–176.
12. Lankard, D. R., 'Applications of fibre concrete,' *Fibre-reinforced Cement and Concrete*, RILEM Symposium 1975, The Construction Press Ltd., pp. 3–19.
13. Luke, C. E. 'Driveway, road, and airport slabs,' *Fibrous Concrete, Construction Material for the Seventies*, Conference Proceedings M-28, CERL, December, 1972, pp. 199–208.
14. Gray, B. H., and Rice, J. L., 'Pavement performance investigation,' *Fibrous Concrete, Construction Material for the Seventies*, Conference Proceedings M-28, CERL, December, 1972, pp. 147–157.
15. Parker, F., 'Construction of fibrous concrete overlay, Tampa International Airport,' *Fibrous Concrete, Construction Material for the Seventies*, Conference Proceedings M-28, CERL, December, 1972, pp. 177–197.
16. State-of-the-art Report on Fibre-reinforced Concrete, *Title No 70–65*, ACI Journal, November 1973, pp. 729–744.
17. Lankard, D. R., and Walker, A. J., 'Pavement and bridge deck overlays with steel-fibrous concrete,' *ACI Special Publication*, 44–22, pp. 375–413.
18. Gregory, J., Galloway, J. W., and Raithby, K. D., 'Full-scale trials of a wire-fibre-reinforced concrete overlay on a motorway,' *Fibre-reinforced Cement and Concrete*, RILEM Symposium 1975, The Construction Press Ltd., pp. 383–394.
19. McDonald, A. R., 'Wirand concrete pavement trials,' *Fibrous Concrete, Construction Material for the Seventies*, Conference Proceedings M-28, CERL, December 1972, pp. 209–234.
20. Kent, B., 'Pavements and overlays,' *Fibrous Concrete: Discussion*, Paper 5,

Concrete Society One-day Symposium Fibrous Concretes USA and UK, September 1972, pp. 49.
21. Swamy, R. N., and Lankard, D. R., 'Some practical applications of steel-fibre reinforced concrete,' *Proc. Institution of Civil Engineers*, Part 1, **56**, pp. 235–256 (1974). August.
22. Johnston, C. D., 'Steel-fibre reinforced concrete pavement—second interim performance report,' *Fibre-reinforced Cement and Concrete*, RILEM Symposium 1975, The Construction Press Ltd., pp. 409–418.
23. Williamson, G. R., 'Fort Hood fibre-concrete overlay,' *Fibre-reinforced Cement and Concrete*, RILEM Symposium 1975. The Construction Press Ltd., pp. 453–459.

Appendix 1
Theory for Minimum Crack Spacing (x') of Long, Non-circular Fibres with Frictional Bond
(See Section 3.4.1)

We can calculate (x') from a simple balance of the load $(\sigma_{mu} V_m)$ needed to break unit area of matrix and the load carried by N fibres across the same area after cracking. This load is transferred over a distance (x') by the limiting maximum shear stress (τ).

For aligned fibres,

$$N = \frac{V_f}{A_f} \tag{A1.1}$$

$$P_f \cdot N \cdot \tau \cdot x' = \sigma_{mu} \cdot V_m \tag{A1.2}$$

where

A_f = cross-sectional area of fibre

and

P_f = perimeter of fibre as shown in Figure A1.1.

From (A1.1) and (A1.2)

$$x' = \frac{V_m}{V_f} \cdot \frac{\sigma_{mu} \cdot A_f}{\tau \cdot P_f} \tag{A1.3}$$

The crack spacing will eventually be between x' and $2x'$.

For film fibres with a rectangular section

$$\frac{A_f}{P_f} = \frac{bt}{2(b+t)} \tag{A1.4}$$

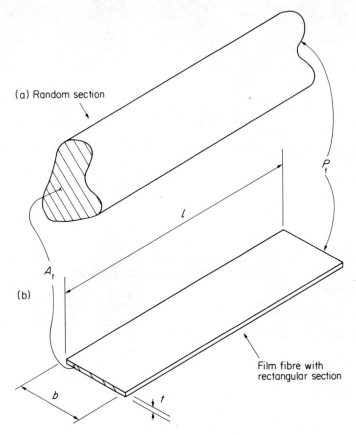

Figure A1.1. Fibre shapes (a) Random section (b) Film fibre with rectangular section

and for situations where $t \ll b$

$$\frac{A_f}{P_f} \simeq \frac{t}{2}$$

Hence for very thin films, equation (A1.3) approximates to

$$x' \simeq \frac{V_m}{V_f} \cdot \frac{\sigma_{mu}}{\tau} \cdot \frac{t}{2} \qquad (A1.5)$$

Appendix 2
Practical Examples using the Theoretical Treatment for Typical Real Composites

1. Determine the change in concrete properties likely to be caused by the inclusion of 0.5 per cent by volume of polypropylene fibres in a 3-D random orientation in concrete.

    Use equation (3.4) and (3.5).

    Assume $E_f = 5 \text{GN/m}^2$ $E_m = 30 \text{ GN/m}^2$,

    Efficiency factor for orientation = 1/5.

2. Assuming that the effective diameter of the fibrillated polypropylene fibres in example 1 is 1.4 mm, calculate the number of fibres which will cross the broken surfaces of a precast component 25 mm thick by 1 m wide.
    Assume 2-D random orientation and use Equation (3.24).

3. A glass reinforced cement composite is made with 2-D random chopped reinforcement. Calculate the cracking stress (i.e. point of non-linearity of the stress/strain curve) assuming a fibre length efficiency factor of 0.6 and an orientation efficiency of 3/8, with a fibre volume of 5 per cent.

    Assume tensile strength of cement matrix = 5 MN/m^2.

    $E_f = 70 \text{ GN/m}^2$ $E_m = 17 \text{ GN/m}^2$.

    Use equation (3.4).

4. A glass-reinforced cement panel is to be made using the spray suction technique with a skin thickness of 10 mm and a fibre volume of 5 per cent.
    Calculate the ultimate tensile strength and post cracking modulus (a) in the longitudinal direction and, (b) in the lateral direction using equation (3.36), (3.37), (3.38), and (3.39).

    Assume $\sigma_f = 1250 \text{ MN/m}^2$ $E_f = 70 \text{ GN/m}^2$.

5. Calculate the modulus of rupture of the glass-reinforced cement material in question 4 in the longitudinal direction assuming that the neutral axis at failure is 2.5 mm from the compression surface.

   Assume the stress block as in Figure 4.2(b).

   Use Equation (4.4).

6. A cracked highway pavement is to be overlayed with 80 mm of steel-fibre concrete with 3 per cent by weight of fibres. The fibres to be used are 0.5 mm dia. x 38 mm long with an average sliding friction bond stress (τ) of 5 MN/m^2.

   Assuming that the overlay cracks, calculate (a) the numbers of wires/mm^2 across the crack and, (b) the average apparent tensile stress which can be maintained across the cracked zone. Compare this with the tensile strength of a plain concrete (3MN/m^2).

   Steel density = 7860 kg/m^3 Concrete density = 2350 kg/m^3.
   Assume 2-D random fibre distribution.
   Use equations (3.24) and (3.30).

7. Assuming that all parameters other than the l/d ratio remain the same as in 6, calculate an appropriate l/d ratio to give a post cracking strength of 3 MN/m^2

   Use equation (3.30).

8. Assuming that the wire parameters remain as in example 6, calculate the percentage of wire by weight to give the composite a post cracking strength equal to 3 MN/m^2.

   Use equation (3.30).

9. For the pavement described in example 6, calculate (a) the average stress in the wire at pullout and, (b) the allowable percentage reduction in wire diameter and cross-sectional area by corrosion, before wire fracture rather than pullout occurs on average.

   Assume wire strength = 1500 MN/m^2.

   Use equations (3.28) and (3.26) respectively.

10. Calculate the modulus of rupture for a fibre concrete, 100 mm in depth of section, with 1.5 per cent by volume of fibres of diameter 0.5 mm and length 40 mm, assuming that the neutral axis is $3D/4$ from the tensile surface: (a) 2-D orientation; (b) 3-D orientation.

    Assume the average sliding friction bond strength
    $\tau = 4.0$ MN/m^2
    Use equations (4.13) and (4.14).

11. Calculate the modulus of rupture for a 500 mm deep beam of the same material as in question 10(b). Assume that failure will occur when the neutral axis is at $0.6D$ from the tensile surface. (It is assumed that, due to the wide primary cracks in this deep beam, the wire pull out force will reduce to zero for neutral axes $> 0.6D$.)

    Assume a rectangular stress block in the tensile zone and a triangular stress block in the compressive zone.

    (NOTE: 1st calculate, σ_{cu} from equation (3.31) and then use the same approach that was used to deduce equation (4.4).)

12. Draw on Figure (4.10) the values of σ_{MR} for $l/d = 60$.

    Use Equation (4.12), (4.13), and (4.14).

13. Calculate the modulus of rupture for the 1 year old glass reinforced cement shown in Figure 4.5 if the strain at failure is 1000×10^{-6}.

    Assume $E = 20$ GN/m^2 $\sigma_{cu} = 10$ MN/m^2

    (Determine the neutral axis position at failure by balancing compressive and tensile areas.)

14. Calculate the minimum crack spacing (x') and crack width (w) at the completion of multiple cracking for 5 per cent by volume of continuous aligned polypropylene film fibres, of film thickness $(t) = 30 \times 10^{-6}$ m.

    Use equations (A1.5) and (3.17).

    Assume $E_f = 5$ GN/m^2 $E_m = 20$ GN/m^2

    $\tau = 1$ MN/m^2 $\sigma_{mu} = 4$ MN/m^2

Solutions to Practical Examples

1. $\sigma_m = \dfrac{\sigma_c}{1 + V_f(M - 1)}$ \hfill (3.4)

   $= \dfrac{\sigma_c}{1 + 0.005 \cdot \frac{1}{5}(\frac{5}{30} - 1)}$

   $= \dfrac{\sigma_c}{0.9992}$

   i.e. The matrix stress is increased by a negligible amount.

   $E_c = E_f V_f + E_m(1 - V_f)$ \hfill (3.5)
   $= 5 \times 10^9 \times 0.005 \cdot \frac{1}{5} + 30(0.995) 10^9$
   $= \underline{29.855 \text{ GN/m}^2}$

   i.e. The modulus is decreased by a negligible amount.

2. $N = \dfrac{2}{\pi} \dfrac{V_f}{\pi r^2}$ per unit area \hfill (3.24)

   $= \dfrac{2}{\pi} \dfrac{0.005}{\pi \times 0.49}$ per mm^2

   $= 0.0021$ fibres per mm^2

   Area $= 25 \times 1000$ mm^2

   $\qquad$ Number of fibres $= \underline{53}$

   NOTE: This is not many, but if the average pull out load of the fibres is known, then the load capacity after cracking can be calculated.

3. Effective fibre volume $V_f = 0.05 \times \frac{3}{8} \times 0.6$

   $\qquad\qquad\qquad = 0.0112$

From equation (3.4)

$\sigma_c = \sigma_m[1 + V_f(M - 1)]$
$= 5 \times 10^6[1 + 0.0112(\frac{70}{17} - 1)]$
$= 5 \times 10^6[1 + 0.0112(3.1)]$
$\simeq \underline{5.2 \text{ MN/m}^2}$

i.e. point of non-linearity is not very different from the failure stress.

4(a). $\sigma_c \simeq 0.27\, \sigma_f\, V_f$ (3.37)
$= 0.27 \times 1250 \times 10^6 \times 0.05$
$= \underline{16.9 \text{ MN/m}^2}$

$E_c \simeq 0.26\, E_f\, V_f$ (3.36)
$= 0.26 \times 70 \times 10^9 \times 0.05$
$= \underline{0.91 \text{ GN/m}^2}$

(b). $\sigma_c \simeq 0.17\, \sigma_f\, V_f$ (3.39)
$= \underline{10.6 \text{ MN/m}^2}$

$E_c \simeq 0.16\, E_f\, V_f$ (3.38)
$= \underline{0.56 \text{ GN/m}^2}$

5. $\sigma_{MR} = 2.44\, \sigma_{cu}$ (4.4)
$= 2.44 \cdot 16.9$
$= \underline{41.2 \text{ MN/m}^2}$

6(a). $N = \dfrac{2}{\pi}\, \dfrac{V_f}{\pi r^2}$ (3.24)

$V_f = \dfrac{0.03}{7860} \times 2350 = 0.009$

$\therefore N = \dfrac{2}{\pi} \times \dfrac{0.009}{\pi(0.25)^2}$

$= \underline{0.029 \text{ wires/mm}^2}$

(b) $\sigma_{cu} = \dfrac{2}{\pi} V_f\, \tau\, \dfrac{l}{d}$ (3.30)

$= \dfrac{2}{\pi} \times 0.009 \times 5 \times 10^6 \times \dfrac{38}{0.5}$

$= \underline{2.18 \text{ MN/m}^2}$

7. $$3 \times 10^6 = \frac{2}{\pi} \times 0.009 \times 5 \times 10^6 \times \frac{l}{d}$$

$$\therefore \quad \frac{l}{d} = \underline{105}$$

8. $$3 \times 10^6 = \frac{2}{\pi} \times V_f \times 5 \times 10^6 \times \frac{38}{0.5}$$

$$\therefore \quad V_f = 0.0124$$

$$\therefore \text{Percentage by weight} = 0.0124 \times \frac{7860}{2350} \times 100$$

$$= \underline{4.15\% \text{ by weight}}$$

9(a) $$\sigma_f = \tau \cdot \frac{l}{d} \tag{3.28}$$

$$= 5.76 = \underline{380 \text{ MN/m}^2}$$

NOTE: This is only about ¼ of normal wire strengths.

(b) Average pullout length $= \dfrac{l}{4}$

Wire pullout force $F = \tau \cdot \pi d \cdot \dfrac{l}{4}$ \hfill (3.26)

$$= 5 \times \pi \times 0.5 \times \frac{38}{4}$$

$$= 74.6 \, N$$

Let corroded diameter $= d_c$

$$\therefore \quad \frac{\pi d_c^2}{4} \times 1500 = 74.6$$

$$\therefore \quad d_c^2 = 0.0633$$

$$\therefore \quad d_c \simeq \underline{0.25 \text{ mm}}$$

∴ The diameter can reduce by about 50 per cent before a change in composite behaviour becomes apparent. Alternatively, about 75 per cent of the cross-sectional area can be rusted away.

10. (a) $\sigma_{MR} = 1.55 \, V_f \, \tau \, \dfrac{l}{d}$

$$= 1.55 \times 0.015 \times 4 \times 10^6 \times \frac{40}{0.5} \tag{4.13}$$

$$= \underline{7.44 \text{ MN/m}^2}$$

(b) $\sigma_{MR} = 1.22 \, V_f \, \tau \, \dfrac{l}{d}$

$$= \underline{5.86 \text{ MN/m}^2} \tag{4.14}$$

11. $\sigma_{cu} = \frac{1}{2} V_f \tau \frac{l}{d}$

$= \underline{2.4 \text{ MN/m}^2}$ (3.31)

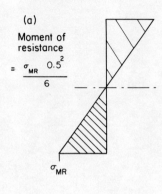
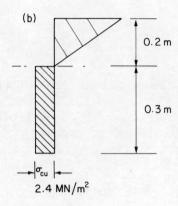

(a) Moment of resistance $= \sigma_{MR} \dfrac{0.5^2}{6}$

σ_{MR}

(b) 0.2 m, 0.3 m

σ_{cu}
2.4 MN/m²

For (b) Moment of resistance $= 2.40 \times 0.3 [0.15 + \frac{2}{3} \times 0.2]$

$= \underline{0.204 \text{ MN m}}$

For (a) Moment of resistance $= \sigma_{MR} \dfrac{0.5^2}{6}$

$= \sigma_{MR} \times 0.042 \text{ MN m}$

∴ $\sigma_{MR} \, 0.042 = 0.204$

∴ $\underline{\sigma_{MR} = 4.86 \text{ MN/m}^2}$

NOTE: Reductions in σ_{MR} from 7.10–5.00 MN/m² for depth increases of 100–250 mm have been observed by Swamy and Stavrides (Chapter 4, reference 16).

12. *Theoretical apparent modules of rupture*

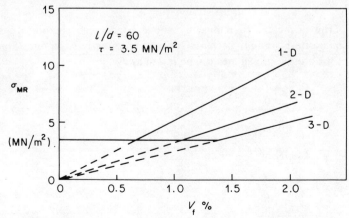

13.

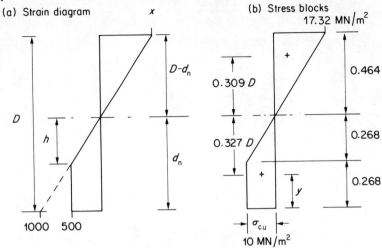

(a) Strain diagram   (b) Stress blocks

Figure (a). Equating areas for stress

$$x \frac{(D - d_n)}{2} = 500(d_n - h) + 500 \frac{h}{2} \tag{A2.1}$$

Similar triangles

$$\frac{500}{h} = \frac{1000}{d_n} = \frac{x}{(D - d_n)} \tag{A2.2}$$

$$\therefore h = 0.5 \, d_n$$

Substituting for h in (A2.1)

$$x(D - d_n) = 1000 \times 0.5 \, d_n + 500 \times 0.5 \, d_n$$
$$\therefore \underline{x(D - d_n) = 750 \, d_n} \tag{A2.3}$$

From (2)

$$x = \frac{1000(D - d_n)}{d_n} \tag{A2.4}$$

Substituting for x in (A2.3)

$$1000(D - d_n)^2 = 750 d_n^2$$

Hence

$$0.25 d_n^2 - 2 d_n D + D^2 = 0$$

Solving for d_n

$$d_n = \frac{+2D \pm \sqrt{(4D^2 - D^2)}}{0.5}$$

$$\therefore \quad d_n = \frac{2D \pm 1.732D}{0.5}$$

$$= 0.536\,D$$

From equation (A2.4)

$$x = 866 \times 10^{-6}$$

Compressive stress

$$= 20 \times 10^9 \times 866 \times 10^{-6} = 17.32 \text{ MN/m}^2 = 1.732\,\sigma_{cu}$$

Figure (b). To find lever arm of stress blocks.

Let distance to centre of gravity of tensile block = y
Take moments about C. of G.

$$y \cdot 10 \cdot \frac{y}{2} = \frac{(0.268 - y)^2}{2} \times 10 + \frac{0.268}{2} \times 10\,(0.089 + 0.268 - y)$$

$$\therefore \quad y^2 = 0.072 + y^2 - 0.536y + 0.096 - 0.268y$$

$$\therefore \quad y = 0.209D$$

$$0.536D - 0.209D = 0.327D$$

$\therefore$ Lever arm

$$= 0.327D + \tfrac{2}{3}0.464D$$

$$= 0.636D$$

$\therefore$ Moments of resistance

$$= \frac{1.732\sigma_{cu} \times 0.464D}{2} \cdot 0.636D$$

$$= 0.256 \times \sigma_{cu} D^2$$

From equation (4.1)

$$\frac{\sigma_{MR} D^2}{6} = 0.256 \sigma_{cu} D^2$$

$$\therefore \quad \sigma_{MR} = 6 \times 2.56 = 15.36 \text{ MN/m}^2$$

14. Equation (A1.5)

$$x' \simeq \frac{V_m}{V_f} \times \frac{\sigma_{mu}}{\tau} \times \frac{t}{2}$$

$$x' \simeq \frac{0.95}{0.05} \times \frac{4}{1} \times \frac{30 \times 10^{-6}}{2} \text{ m}$$

$$\underline{x' \simeq 1.14 \text{ mm}}$$

Equation (3.17)

$$w = \epsilon_{mu}(1 + \alpha)x'$$

where $\alpha = \dfrac{E_m V_m}{E_f V_f}$

$$w = \frac{4 \times 10^6}{20 \times 10^9} \left(1 + \frac{20 \times 0.95}{5 \times 0.05}\right) 1.14$$

$\therefore \quad \underline{w = 0.018 \text{ mm}}$

Author Index

Adams, M. A. J., 60
Addington-Smith, T. D., 180
Al-Hassani, S. T. S., 97
Ali, M. A., 45, 46, 51, 117, 118, 119, 123, 125, 133, 148, 152, 155
Al-Kayyali, O. A., 87, 97
Allen, H. G., 12, 19, 24, 33, 38, 51, 106, 109, 110, 122, 133, 140, 142, 145, 171, 180
Al-Noori, K. A., 78
Argon, A. S., 33
Aveston, J., 16, 17, 18, 19, 20, 21, 22, 31, 33, 37, 40, 41, 51, 73, 74, 79, 148, 149, 150, 151, 156

Bailey, J. E., 146, 155
Bailey, J. H., 79
Bailey, L., 78
Bailey, M. B., 156
Ball, C., 78
Barab, S., 180
Barker, H. A., 155
Batson, G. B., 31, 34, 78
Beckett, R. E., 79
Bekaerts, N. V. Ltd., 60
Bentley, S., 79
Bergström, S. G., 60, 78
Bills, P. M., 136
Biryukovich, D. L., K. L., and Yu. L., 99, 101, 106, 107, 109, 110, 133, 180
Blood, G. W., 79
Bowen, D. H., 156
Briggs, A., 150, 152, 156
Brown, J. H., 179, 180
Building Research Establishment, 51

Cappacio, G., 97, 156
Chan, H. C., 110, 133

Clarke, L. L., 128, 134
Cohen, E. B., 115, 116, 134
Coleman, R. A., 32, 34, 78
Cooper, G. A., 31, 32, 51, 73, 79, 156
Cox, H. L., 12, 33
Crompton, P. A., 97
Croney, D., 183, 196
Czernin, W., 2

Dardare, J., 97
Dave, N. J., 175, 181
Davis, H. E., 2
Diamond, S., 116, 134
Dixon, J., 79
Dupont, 3

Edgington, J., 32, 34, 38, 51, 54, 55, 59, 60, 64, 66, 67, 71, 73, 75, 76, 77, 78, 79
Elvery, R. H., 63, 78

Fairweather, A. D., 95, 98, 164, 180
Fattuhi, N. I., 33, 97

Galloway, J. W., 60, 196
Gilson, J. C., 137, 145
Goldfein, S., 96, 153, 156
Gray, B. H., 60, 196
Gregory, J., 60, 196
Grimer, F. J., 126, 133, 153, 156

Hannant, D. J., 32, 33, 34, 37, 38, 43, 47, 48, 49, 51, 54, 55, 60, 64, 66, 73, 75, 76, 78, 79, 80, 90, 91, 92, 97, 98, 180, 184, 196
Hanson, D., 180
Haynes, B. C., 181
Hearle, J. W. S., 97
Hejgaard, O., 134, 154, 156
Henry, R. L., 163, 180

Hibbert, A. P., 79, 98, 126, 132, 133, 134, 153, 156
Hills, D. L., 127, 133
Hobbs, C., 88, 97, 180
Hodgson, A. A., 145
Hoff, G. C., 69, 78, 158, 179
Holiday, L., 33
Holister, G. S., 33
Hooks, J., 78
Hughes, B. P., 33, 89, 97
Hughes, D. C., 90, 91, 92, 97

Ingerslev, E., 180

Jaras, A. C., 134
Johnson and Nephew (Ambergate) Ltd., 60, 79
Johnston, C. D., 32, 34, 43, 48, 51, 60, 63, 66, 69, 72, 78, 193, 197
Jones, F. E., 143, 145

Kaden, R. A., 158, 179
Kakimi, N., 79, 180
Kelly, A., xi, 31, 33, 34, 51, 73, 79, 90, 97, 156
Kelly, J. W., 2
Kent, B., 60, 192, 196
Key, W. H., 30, 34
Klos, H. G., 135, 137, 140, 145
Kollek, J., 156
Krenchel, H., 12, 29, 31, 32, 33, 34, 51, 100, 101, 130, 133, 134, 145, 152, 153, 154, 155

Landers, E., 78
Lankard, D. R., 66, 78, 79, 157, 163, 179, 180, 189, 192, 194, 196, 197
Larner, L. J., 115, 116, 133, 134, 144, 145
Lawrence, P., 134
Laws, V., 12, 25, 26, 28, 33, 45, 46, 51, 134
Litherland, K. L., 134
Loe, J. A., 196
Luke, C. E., 196

McCabe, P. J., 68, 79
McCurrich, L. H., 60
McDonald, A. R., 60, 192, 196
Mackintosh, D. M., 97
Majumdar, A. J., 28, 33, 51, 95, 97, 99, 108, 110, 111, 112, 115, 116, 117, 118, 119, 123, 125, 131, 133, 134, 143, 144, 145, 147, 148, 152, 153, 155, 156, 180

Malhotra, V. M., 145
Mangat, P. S., 33, 56, 60, 78
Marsh, H. N., 128, 134, 181
Mayfield, B., 78, 79
Mercer, R. A., 17, 18, 22, 33, 37, 40, 41, 51, 74, 79, 149, 150, 151, 156
Moens, J., 51
Monfore, G. E., 156

Naaman, A. E., 33, 62, 78
Nanda, V. K., 97
Nawy, E. G., 180
Neville, A. M., 2
Newserth, G. E., 180
Nishioka, K., 69, 79, 163, 180
Nurse, R. W., 33, 51, 99, 110, 111, 131, 133, 134

Oakley, D. R., 28, 33, 51, 101, 133
O'Leary, D. C., 181
Opoczky, L., 143, 145
Orchard, D. F., 2

Parker, F., 60, 196
Patterson, W. A., 110, 133
Pecuil, T. E., 181
Pell, P. S., 79
Pennings, A. J., 97
Pentek, L., 143, 145
Phillips, J., 180
Pilkington Brothers Ltd., 99
Pomeroy, C. D., 170, 179, 180
Proctor, B. A., 28, 33, 51, 101, 133

Rahman, T. A., 97
Raithby, K. D., 60, 196
Ramey, M. R., 68, 79
Rangan, B. V., 32, 34, 78
Rao, C. V. S. K., 78
Rao, S. V. K., 33
Raouf, Z. A., 97
Rayment, D. L., 147, 148, 155
Rice, J. L., 60, 184, 185, 196
Ritchie, A. G. B., 87, 88, 97
Roberts, N. P., 79
Romualdi, J. P., 31, 34
Ryder, J. F., 134, 145, 180, 181

Salmons, R. F., 96
Samarai, M. A., 78
Samuels, R. J., 85, 97
Sarkar, S., 156
Saunders, J., 181
Shack, W. J., 33

Shah, S. P., 30, 32, 33, 34, 62, 78
Sheets, H. D., 79, 180
Shirakawa, K., 180
Sillwood, J. M., 17, 18, 22, 33, 37, 40, 41, 51, 74, 79, 149, 150, 151, 156
Simons, J. W., 181
Simpson, J. W., 97
Singh, B., 117, 118, 119, 123, 125, 133
Soames, N. F., 180
Soare, A. J. M., 173, 180
Speakman, K., 133, 134
Spencer-Smith, J. L., 82, 96
Spring, N., 43, 51, 60, 78
Stavrides, H., 51, 56, 60, 64, 65, 67, 78, 206
Stucke, M. S., 51, 112, 133
Swamy, R. N., 33, 51, 56, 60, 64, 65, 66, 67, 69, 78, 79, 192, 197, 206
Swift, D. G., 155
Snyder, J. M., 78
Szabo, H., 180

Tattersall, G. H., 77
Thomas, C., 33
Troxell, G. E., 2

Urbanowicz, C. R., 77, 155
Uzomaka, O. J., 146, 155

Walker, A. J., 157, 179, 189, 196
Waller, J. A., 150, 156
Walton, P. L., 95, 97, 128, 133, 153, 156, 180
Ward, I. M., 97, 156
West, J. M., 116, 134, 144, 145
Westergaard, H. M., 183, 184, 196
Williams, J. R., 173, 180
Williams, R. I. T., 32, 34, 51, 54, 55, 60, 66, 73, 78, 157, 179, 182, 196
Williamson, G. R., 69, 79, 153, 156, 194, 197
Winer, A., 145

Yamakawa, S., 79, 180

Zelly, B., 78
Zia, P., 79
Zonsveld, J. J., 81, 84, 89, 90, 91, 92, 96, 97, 180
Zweben, C., 90, 97
Zwijnenburg, A., 97

Subject Index

A-glass, 100, 101
Air entrained concrete, polypropylene
  fibre, 88
  steel fibre, 56, 57
Airfield pavements, 182, 185, 188
Akwara fibres, 146
Aligned fibres, glass, 106, 107
  polypropylene, 90, 92
  steel, 67
  theory, 15, 16, 21, 41
Alkali resistant glass fibres, 99–132
  applications, 172
  composition, 100
  corrosion, 116
  properties, 4, 118
Alumina fibres, 146
Asbestos, 135
  amphibole group, 136
  bond strength, 140
  chrysotile, 135
  corrosion, 143
  crocidolite, 136
  health hazards, 137
  in concrete, 145
  properties, 4
  tensile strength, 4, 144
  world production, 178
Asbestos cement, 137–145
  applications, 177
  compressive strength, 143
  corrugated sheets, 178
  density, 143
  durability, 143
  fibre volume, 137, 143
  flexural strength, 140, 143
  health hazards, 179
  impact resistance, 143
  modulus of elasticity, 141, 143
  pressure pipes, 139, 178
  production technology, 137, 179
  products in U.K., 178
  tensile strength, 141, 143
  tensile stress–strain behaviour, 19, 20, 141, 142
  thermal expansion, 143
  void content, 141
  water absorption, 143
Aspect ratio, effect on flexural properties, 42, 49, 72
  effect on workability, 54

Bond, effect on, crack spacing, 17, 21
  critical fibre volume, 14
  efficiency factors, 26
  failure strain of composite, 43
  flexural strength, 43, 49
  tensile strength, 24

Carbon fibre, properties, 4, 146, 147
Carbon fibre cement, creep, 150
  durability, 148, 152
  fatigue life, 151, 152
  flexural strength, 148, 150
  impact strength, 147, 148
  modulus of elasticity, 147, 151
  shrinkage, 150
  tensile stress–strain behaviour, 147, 148, 149, 150
Carbonation, at cracks, 75, 76
  of asbestos fibres, 143
Caricrete, 81, 164, 165
Cavitation, 69, 158
Cellulose fibre, 4, 152
Cem-FIL, 171
Charpy test, 69, 70, 71
Chrysotile asbestos, 135
Coconut fibres, 152
Compacting factor, 53, 87

Corrosion of, alkali resistant glass, 116
   asbestos fibres, 143, 144
   E-glass, 115
   steel wires, 75, 76
Corrugated sheet roofing, 171, 178
Cotton, 82
Crack, first, 9
   spacing, 17, 18, 21, 22, 25, 30, 199
   suppression, 19, 20
   width, 16, 18, 63
Cracking, in composite, 11, 31
   in slabs, 63
   in steel fibre concrete, 63
   multiple, 14, 16, 19, 20, 90, 91, 93
Critical fibre, length, 12, 21
   volume, 13, 14, 15, 25, 27, 28, 29, 39, 42
Crocidolite asbestos, 136

Denier, 84
Dolosse breakwater units, 161
Draw ratio, 81, 82
Ductility in tension, effect on flexural strength, 43, 50, 112
Durability, asbestos cement, 143
   carbon fibre cement, 148, 152
   glass fibre cement, 109, 110, 112, 115, 116, 121, 122, 124, 125, 171
   polypropylene fibres, 95, 96
   steel fibre concrete, 74, 75, 76, 77

Efficiency factors, length, 24
   post-cracking, 22, 23, 25, 26, 29
   pre-cracking, 12
Elastic modulus, *see* Modulus of elasticity
Elkalite, 106, 171
Energy absorption, 71
Energy costs, xi

Faircrete, 86, 88, 169
Fatigue strength, carbon fibre cement, 151
   glass fibre cement, 126
   steel fibre concrete, 68
Fibre, balls, 58
   cost, 7
   distribution, effect of aggregate size, 55
   number crossing a unit area, 23
   orientation, effect on flexural strength of steel fibre concrete, 43, 49, 59, 64

orientation, effect on properties of asbestos cement, 141, 143
orientation, for theoretical effects, *see* Efficiency factors
properties, 4
quantity, equation for workability of steel fibre concrete, 58
quantity, for batching, 6
spacing, 32, 33
spacing, effect on composite strength 32, 46, 47
stress at pull-out, 24, 27
stress distribution at a crack, 18
stress transfer length, 17
volume for flexural strengthening, 42
Fibrillation, 83
Filler, 5
Fire resistance, asbestos cement, 178
   glass fibre gypsum, 130, 177
   Kevlar cement, 153
   polypropylene concrete, 95
   steel fibre concrete, 69, 163
Flexural strength (modulus of rupture), for different composites, *see under* fibre type
   theory, 35–51
Flexure toughness, 72
Flotation units, 166, 167
Fly ash, 5, 56, 57, 102, 195
Formwork, asbestos cement, 178
   glass reinforced cement, 175
Fracture mechanics, 31
Freeze–thaw behaviour, 75, 152

Glass fibre, bond strength, 118
   composition of A-glass, E-glass, and Zirconia glass, 100
   corrosion of alkali-resistant, 116
   corrosion of E-glass, 115
   critical volume, 15, 28
   distribution in matrix, 132
   durability, 112
   efficiency factors, 28, 29
   manufacture, 99, 100
   properties, 4, 116, 118
   strand, cross-sectional area, 101, 108
   strand, perimeter, 101, 108
Glass fibre in concrete, 128
   applications, 176, 177
   flexural strength, 129
Glass fibres in gypsum plaster, 130
   applications, 177
   flexural strength, 131

Glass fibres in high alumina cement, 106–111
  applications, 171
  compressive strength, 109
  flexural strength, 109, 111
  impact strength, 110
  tensile stress–strain behaviour, 106, 107, 108
Glass fibres in ordinary Portland cement, 111–128
  applications, 171–176
  compressive strength, 127
  cracking stress, 11
  density, 104, 118
  example calculations, 11, 15, 28, 29, 200–209
  fatigue strength, 124
  flexural properties, 121, 122, 123
  flexure, effect of age and storage conditions, 122–125
  flexure, effect of ductility, 50
  flexure, stress distribution, 36, 37
  fracture surface (interfacial region), 112–115
  impact strength, 123, 125
  modulus of elasticity, initial, 118, 121
  modulus of elasticity, post-cracking, 29
  plus polypropylene fibres, 128
  pre-mixed, 127
  production techniques, 101
  surface coatings, 176
  tensile strength, effect of fibre volume and fibre length, 117
  tensile stress–strain behaviour, 19, 20, 119
Glass wool, 99, 130
Gunite, polypropylene mortar, 167, 168
  steel fibre concrete, 59, 158, 159, 160
Gypsum plaster, 130, 131

Hatschek process, 138, 139
Health and Safety at Work Act 1974, 179
High alumina cement, properties, 5
  with glass fibres, 106
Highway pavements, 157, 182–197
Hydraulic structures, 158

Impact testing, 7
  for impact resistance of different fibre composites, *see under* fibre type

Interfacial region, glass fibre cement, 111–115
Isotactic polypropylene, 81, 84

Kevlar fibres, 4, 152, 153

Lay-up process, glass reinforced cement, 104
  polypropylene cement, 87
Light transmission technique, 132
Load Classification Number (LCN), 182

Magnani process, 139
Magnetic orientation, steel fibres, 59
Manhole covers, steel fibre concrete, 159
Manholes, polypropylene concrete, 170
Manville extrusion process, 139
Marine applications, glass fibre cement, 176
  polypropylene concrete, 166, 167
  steel fibre concrete, 161
Matrix cracking, 9, 11
  microstructure, 111–115
  properties, 5
Mazza process, 139
Melt extract process, 61
Mesothelioma, 137
Methyl cellulose, 102
Microcracks, 9, 130, 153
Mix design, glass fibre cement, 101, 102
  glass fibre concrete, 128
  polypropylene concrete, 85, 86, 88
  steel fibre concrete, 54–58
Modulus of elasticity, of composite, 11
  of fibre, 4
  of matrix, 5, 73, 118
  for different composites, *see under* fibre type
Modulus of rupture, theoretical principles, 35–37, 40–45 (*see also* Flexural strength)
Moment of resistance, 38
Mortar, properties, 5, 73

Networks of fibrillated film, 5, 83, 84, 87, 91, 92, 93
Neutral axis, position, 36, 38, 109
Nylon fibre, 4, 153

Packerhead process, 163
Pavements, steel fibre concrete, 57, 157, 182–197
  overlays, design, 185, 186, 187

overlays, glass fibre concrete, 176
overlays, steel fibre concrete, 57, 157, 182–197
Perlon fibre, 153, 154
Piassave fibre, 146
Pile shells, polypropylene concrete, 95, 164, 165
Piling, asbestos cement sheets, 178
   glass fibre sheets, 176
Pipes, asbestos cement, 139, 179
   British Standards, 96
   glass fibre concrete, 177
   steel fibre concrete, 163
Plaster of Paris, 131
Poisson's ratio of fibres, 4
   effect on debonding, 90, 91
Polyacrylonitrile, 147
Polyethylene fibres, 155
Polyethylene oxide, 102, 139
Polyolefins, increased crystallization, 83
Polypropylene fibres, bond strength, 89, 90
   critical volume, 16, 29
   durability, 95, 96
   manufacture, 81, 82, 83, 84
   modulus of elasticity, 4, 85
   properties, 4, 82, 83, 84, 85
   ropes, 170
Polypropylene fibre concrete (chopped fibres), applications, 164–170
   compressive strength, 92
   example calculations, 16, 200, 203
   fire resistance, 84, 95
   flexural strength, 92, 94
   impact resistance, 94, 95
   mixing, 85, 86
   stress–strain behaviour, 20
   thixotropy, 88
   workability, 87, 88
Polypropylene film networks in mortar, applications, 170
   flexural strength, 92, 93
   stress–strain behaviour, 91
Post-cracking behaviour, flexural theory, 37–50
   tensile theory, 16–29
Precast flags, British Standards, 96
Pulverized fuel ash, 5, 56, 57, 102, 195

Rayon, 147
Reflection cracking, 186
Refractory applications, 69, 163
Rock wool fibre, 155

Serpentines, asbestos, 135
Sewer linings, 175
Shotcrete, *see* Gunite,
Sisal fibres, 155
Slipform paver, 189, 190, 191, 193
Slump test, 52, 87
Specific fibre surface, 29, 30, 130, 153, 154
Spherulites, 82
Spray-suction process, glass fibre cement, 103
Steel fibre, bond strength, 27
   corrosion, 75, 76, 77
   critical volume, flexure, 47, 48
   critical volume, tension, 15, 25, 26, 27
   magnetic alignment, 59, 64
   manufacture, 61
   melt extract, 61
   properties, 4
   shapes, 62
   stainless, 4, 61, 69, 163
   surface treatment, 62
   tensile strength, 4, 62
Steel fibre concrete, applications, general, 157–164
   applications, pavements and pavement overlays, 182–197
   compaction techniques, 58
   compressive strength, 68
   cracking stress, 11, 63
   creep, 74
   design of pavement overlays, 185, 186, 187
   design of pavements for new construction, 184, 185
   durability, 75, 76, 77
   dynamic strength, 69, 70, 71
   example calculations, 11, 15, 25, 26, 27, 201–206
   fatigue strength, 68
   flexural strength (modulus of rupture), effect of,
      bond strength, 49
      casting direction, 64, 65
      fibre length/diameter ratio, 47, 48, 49
      fibre orientation, 49, 64
      fibre volume, 47, 48, 49, 66
      specimen dimensions, 67
      theoretical, 49
   impact strength, 69, 70, 71
   mix design, 56, 57, 58, 183, 184

mixing methods, 58
modulus of elasticity, 11, 73
shrinkage, 74
sprayed concrete, 59
tensile strength, 32, 65, 66
tensile stress—strain behaviour, 20, 39, 73, 74
torsional strength, 66, 68
workability, 52—58
Straw, xi
Stress block, 36—40, 44, 45
Stress distribution at a crack, 18
Stress—strain behaviour of composites, *see under* fibre type.
Superplasticizers, 56

Tensile strength, *see under* fibre type
Test methods, 7
Thixotropy, 88
Torsional strength, 66, 68

Ultraviolet radiation, 85

Vane test, 88
V—B Consistometer, 53, 54, 87
Vegetable fibres, 82, 146, 152, 155

Westergaard analysis, 183, 184, 194
Workability, compacting factor test, 53, 87
  effect of aggregate size and volume, 54, 55
  effect of fibre length and diameter, 53, 54
  of glass fibre concrete, 128, 129
  of polypropylene fibre concrete, 87, 88
  of steel fibre concrete, 52—58
  slump test, 52, 87
  V—B consistometer test, 53, 54, 87
Woven meshes, 5

Zirconia glass, 100